# Comprehensive Gynecology and Obstetrics

**Series Editors**
Ikuo Konishi, National Kyoto Medical Center
Kyoto, Kyoto, Japan

Hidetaka Katabuchi, Department of Obstetrics and Gynecology
Kumamoto University
Kumamoto, Kumamoto, Japan

This series presents the current and future perspectives of medical science in gynecology and obstetrics. The authors fully describe the current understanding of a disease including clinical features, imaging, pathology, and molecular biology, and also include the historical aspects and theories for exploring the etiology of the disease. Also, recent developments in diagnostic strategy, medical treatment, surgery, radiotherapy, prevention, and better health-care methods are clearly shown. Thus, each volume in the series focuses on the scientific basis for the pathogenesis of a disease and provides clinical applications that make it possible to offer personalized treatment for each patient. Over the past 20 years, physicians have been working to develop a standard treatment and publish clinical guidelines for a disease based on epidemiological evidence, mainly through the use of randomized clinical trials and meta-analyses. Recently, however, comprehensive genomic and genetic analyses have revealed the differences and variations in biological characteristics even among patients with the same diagnosis and have been focusing on personalized therapy. Now all physicians and patients are entering a new world of "precision medicine" through the use of genomic evidence. We are confident that readers will greatly benefit from the contents of the series with its purview of the exciting and promising future of gynecology and obstetrics.

Masaki Mandai
Editor

# Personalization in Gynecologic Oncology

*Editor*
Masaki Mandai
Department of Gynecology and Obstetrics
Graduate School of Med., Kyoto Univ.
Kyoto, Japan

ISSN 2364-1932 ISSN 2364-219X (electronic)
Comprehensive Gynecology and Obstetrics

ISBN 978-981-19-4713-1 ISBN 978-981-19-4711-7 (eBook)
https://doi.org/10.1007/978-981-19-4711-7



This Springer imprint is published by the registered company Springer Nature Singapore Pte Ltd.
The registered company address is: 152 Beach Road, #21-01/04 Gateway East, Singapore 189721, Singapore

# Contents

# Chapter 1
# Personalized Treatment in Ovarian Cancer

**Nozomu Yanaihara, Yasushi Iida, Masataka Takenaka, Ayako Kawabata, Takafumi Kuroda, and Aikou Okamoto**

**Abstract** Ovarian cancer (OC) is a deadly gynecological malignancy, as the majority of patients are diagnosed at advanced stages. In recent years, the development and clinical application of molecularly targeted therapies such as bevacizumab (Bev), poly ADP-ribose polymerase (PARP) inhibitors, and anti-programmed cell death (PD)-1 antibodies may offer potentially beneficial treatment options for patients with advanced OC. However, the clinical utility of these therapies remains limited because they are associated with specific molecular abnormalities. Therefore, the identification of reliable molecular biomarkers is necessary for appropriate patient selection for these new treatment strategies. In this review, we outline the rationale for the selection of novel targets in advanced OC and discuss new therapeutic strategies that can be positioned as precision medicine for patients with advanced OC based on their broad mechanisms of action.

**Keywords** Angiogenesis inhibitors · Antibody–drug conjugates · Immune checkpoint inhibitors · Ovarian cancer · PARP inhibitors · Personalized treatment

## 1.1 Introduction

Ovarian cancer (OC) is a fatal gynecological malignancy that accounts for 5–6% of all cancer-related deaths. OC generally shows few specific symptoms, and the majority of patients are diagnosed at advanced stages. The standard treatment for OC has been primary debulking surgery with the goal of no grossly visible residual tumor, followed by platinum-taxane chemotherapy. Although the "one-size-fits-all"

---

N. Yanaihara (✉) · Y. Iida · M. Takenaka · A. Kawabata · T. Kuroda · A. Okamoto
Department of Obstetrics and Gynecology, The Jikei University School of Medicine, Tokyo, Japan
e-mail: yanazou@jikei.ac.jp



M. Mandai (ed.), *Personalization in Gynecologic Oncology*, Comprehensive Gynecology and Obstetrics, https://doi.org/10.1007/978-981-19-4711-7_1

approach to the treatment of OC has not yielded satisfactory results, recent significant advances in the development and clinical application of targeted therapies may provide beneficial treatment options for a substantial population of patients with OC. Namely, molecularly targeted therapies such as bevacizumab (Bev), poly ADP-ribose polymerase (PARP) inhibitors, and anti-programmed cell death (PD)-1 antibodies are helping to overcome the poor prognosis of OC. Since the clinical efficacy of these therapies is thought to be associated with specific molecular abnormalities, the identification of reliable biomarkers by molecular testing is necessary for clinical application. The development of these molecular targeted therapies has garnered attention as a new treatment for OC, but their efficacy is still limited and mortality in advanced cases remains substantially problematic. This review outlines the rationale for the selection of novel targets in OC and discusses new treatment strategies that can be positioned as precision medicine for patients with advanced OC based on their broad mechanisms of action.

## 1.2 Angiogenesis Inhibitors

Angiogenesis is essential for survival and progression in various tumor types, including advanced OC. Vascular endothelial growth factor (VEGF) is a protein signaling molecule that can stimulate endothelial cells to promote angiogenesis. A recombinant humanized monoclonal antibody against VEGF-A, Bev, is the most widely studied antiangiogenesis agent for OC.

In 2011, the results of GOG-0218 and ICON-7, randomized phase III trials that added Bev to frontline chemotherapy for advanced OC, were published simultaneously [1, 2]. Although the duration and dose of Bev were different in the two studies, both showed a significant increase in progression-free survival (PFS). When compared with the control group, patients who received chemotherapy alone, the median PFS of the Bev group, patients who received Bev as upfront and maintenance therapy, was greater by 3.8 months (14.1 vs. 10.3 months; hazard ratio [HR]: 0.71; 95% CI: 0.63–0.82) in GOG-0218 and 1.5 months (21.8 vs. 20.3 months; HR: 0.81; 95% CI 0.70–0.94) in ICON-7, respectively. In the final analysis of the two studies, there was no significant difference in overall survival (OS) between the Bev group and the control group in the overall population, but in the subgroup analysis of high-risk patients, OS was increased in the Bev group compared with the control group [3, 4]. In GOG-0218, in the subgroup patients with stage IV, the median OS for the Bev group was 42.8 vs. 32.6 months for the control group (HR: 0.75; 95% CI: 0.59–0.95). In ICON7, in the subgroup of patients with high risk of progression defined as stage IV, inoperable stage III, or suboptimally debulked (>1 cm) stage III, median OS for the Bev group was 39.7 vs. 30.2 months for the control group (HR: 0.78; 95% CI: 0.63–0.97). Regarding histological subgroup analysis, patients with clear cell carcinoma (CCC), stage I-IIA CCC or grade 3 carcinoma, and low-grade serous did not benefit from the addition of Bev on OS in ICON7.

The randomized phase III trials evaluating the efficacy of Bev combined with gemcitabine and carboplatin (GC) for patients with platinum-sensitive recurrent OC (OCEANS) showed that the median PFS of GC + Bev was increased by 4.0 months compared with the GC + placebo (12.4 vs. 8.4 months; HR: 0.484; 95%CI: 0.388–0.605) [5]. However, the final OS analysis of OCEANS revealed that there was no significant difference in OS between the groups (33.6 vs. 32.9 months; HR: 0.95; 95% CI: 0.77–1.18) [6]. In GOG-0213, patients with platinum-sensitive recurrent OC were randomized to receive standard chemotherapy with paclitaxel and carboplatin (TC) or TC + Bev with Bev maintenance therapy [7]. Although the median OS of the Bev group was increased by 4.9 months compared with the TC group (42.2 vs. 37.3 months), the difference was not statistically significant (HR: 0.829; 95% CI: 0.683–1.005). Additionally, a randomized phase III study (AGO-OVAR 2.21) comparing PFS of control group as GC + Bev with that of carboplatin and pegylated liposomal doxorubicin (PLD) (PLD-C) + Bev showed that the median PFS and OS of experimental group was prolonged by 1.7 and 4.1 months compared with control group, respectively (PFS: 13.3 months vs. 11.6 months; HR:0.81; 95% CI 0.68–0.96) (OS: 31.9 months vs. 27.8 months; HR: 0.81; 95% CI: 0.67–0.98) [8]. Recently, a randomized phase III trial (MITO16b/MANGO-OV2) aiming to reveal the benefit of Bev in combination with chemotherapy for patients with recurrent OC who have already received bevacizumab in first-line therapy showed that the median PFS of chemotherapy with Bev group and control group was 11.8 and 8.8 months, respectively (HR: 0.51; 95% CI: 0.41–0.65) [9]. For patients with platinum-sensitive recurrent OC, the addition of Bev to platinum-based chemotherapy doublets, especially PLD-C, might be considered in clinical practice, even if Bev has already been used in first-line therapy.

In the randomized phase III trial (AURELIA) designed to determine the impact of Bev with chemotherapy for patients with platinum-resistant recurrent OC who had received no more than two prior lines of chemotherapy, patients were randomly assigned to control group, patients who received investigator-chosen single-agent chemotherapy (PLD, paclitaxel, topotecan) or Bev group, patients who received single-agent chemotherapy with Bev. The median PFS of Bev group and control group was 6.7 and 3.4 months, respectively (HR: 0.42; 95% CI: 0.32–0.53). However, there was no significant difference in OS between the two groups [10].

In order to identify the predictive biomarkers for selecting candidates for Bev, several studies have been conducted using samples from clinical trials. Five tumor biomarkers (VEGF-A, VEGFR-2, neuropilin-1, MET, CD31) were assessed by immunohistochemistry in GOG-0218. The effect of Bev on both PFS and OS was associated with high microvessel density measured by CD31. VEGF-A was not a predictor of PFS, but using the third-quartile cutoff for high VEGF-A expression showed that it could be a predictor of OS [11]. In addition, plasma samples from GOG-0218 were analyzed for seven biomarkers (IL6, Ang-2, osteopontin (OPN), stromal cell-derived factor-1 (SDF-1), VEGF-D, IL6-R, and GP130) that had been previously shown to be predictive of benefit from anti-angiogenetic therapies in other solid tumors and/or associated with outcomes with OC. Among them, IL6 was predictive of a therapeutic benefit with Bev for PFS and OS [12]. In ICON-7, three candidate biomarkers including mesothelin, fms-like tyrosine kinase-3 (FLT4), and $\alpha_1$-acid glycoprotein (AGP) were identified by analyzing serum samples from 10 patients who received Bev (five responders and five nonresponders). When combined with CA125, those biomarkers identified therapeutic benefits with Bev for PFS [13]. Homologous recombination (HR) repair mutations, including *BRCA1* and *BRCA2*, significantly prolonged PFS and OS in GOG-0218 cohort, had no correlation with response to Bev [14]. Imaging biomarkers of Bev have been investigated. Ascites are a feature of advanced OC, and VEGF expression has been suggested to be involved in the pathogenesis of ascites. In the study using data from GOG-0218, patients with ascites treated with Bev had longer PFS and OS [15]. Adiposity has been hypothesized to interfere with the activity of Bev, so several studies had been conducted to determine whether makers of adiposity are a potential clinical biomarker for the efficacy of Bev. In a study, pretreatment computed tomography (CT) for GOG-0218 participants were analyzed, and markers of adiposity, including BMI, surface fat area (SFA) and visceral fat area (VFA), were not predictive of benefit with Bev [16]. In another study using CT performed after primary debulking surgery for GOG-0218 participants, subcutaneous fat density (SFD) and visceral fat density (VFD) were used as markers of adiposity, and median OS was prolonged with Bev compared to placebo in the VFD high group [17]. There are no prospectively established biomarkers for response to Bev, so the identification of such predictive biomarkers remains an urgent unmet clinical need in OC treatment.

## 1.3 PARP Inhibitors

PARP inhibitors result in an accumulation of DNA damage by multiple mechanisms, including synthetic lethality and PARP trapping, leading to cell death which has HR deficiency (HRD), such as cancer cells with pathogenic variants of *BRCA1/2* genes [18]. The appearance of the PARP inhibitors has changed the standard care of OC and dramatically improved survival in patients with specific biomarkers, such as *BRCA1/2* mutations and HRD. Recent well-conducted clinical trials have shown the efficacy of PARP inhibitors as maintenance treatment following chemotherapy in the setting of both newly diagnosed and recurrent OC.

In the first-line setting of advanced OC, four well-conducted clinical trials, SOLO1, PAOLA-1, PRIMA, and VELIA trials, have shown the clinical benefit of PARP inhibitors (Table 1.1). The SOLO1 study demonstrated the efficacy of olaparib as first-line maintenance therapy in stage III and IV high-grade serous ovarian cancer (HGSOC) and endometrioid ovarian cancer (EMOC) with germline *BRCA* (g*BRCA*) mutation [19]. Olaparib maintenance therapy was performed in tumors of partial or complete response to prior chemotherapy. After the 5-years follow-up from randomization, olaparib maintenance therapy significantly improved PFS compared with administered placebo (HR: 0.33; 95% CI: 0.25–0.43) [20]. The study was a first phase III study that showed survival benefit of maintenance olaparib in OC with g*BRCA* mutation in the first-line setting. In a subgroup analysis by

**Table 1.1** Representative clinical trials with PARPi in advanced and recurrent ovarian cancer

| | Patients | No. | Intervention | Stratification | Biomarkers | Population | Median PFS (months): HR (95% CI) |
|---|---|---|---|---|---|---|---|
| *Maintenance therapy in the first-line setting* | | | | | | | |
| SOLO1 | • Newly diagnosed ovarian, peritoneal, fallopian-tubal cancer<br>• Stage III or IV<br>• High-grade serous or endometrioid histologies<br>• Having germline or somatic *BRCA1* or *BRCA2* mutation<br>• NED or a clinical CR/PR to platinum-based chemotherapy | 391 | • Maintenance olaparib (260)<br>vs.<br>• Maintenance placebo (131)<br>Until PD (up to 2 years) | • Clinical response after platinum-based chemotherapy (CR or PR) | BRC Analysis test (Myriad)<br>*BRCA*1/2 genetic testing assay (BGI) | All patients (g*BRCA* mt) | 56.0 vs. 13.8: 0.33 (0.25-0.43) (higher clinical risk)<br>40.6 vs. 11.1: 0.34 (0.24-0.49) (lower clinical risk)<br>NR vs. 21.9: 0.38 (0.25–0.59) |
| PAOLA-1 | • Newly diagnosed ovarian, peritoneal, fallopian-tubal cancer<br>• Stage III or IV<br>• High-grade serous, endometrioid, or other histologies<br>• NED or a clinical CR/PR to platinum-taxane chemotherapy plus bevacizumab | 806 | • Maintenance olaparib and bevacizumab (537)<br>vs.<br>• Maintenance placebo and bebvacizumab (269)<br>Until PD (up to 2 years) | • Outcome of first-line treatment at screening<br>• Tumor *BRCA* status | myChoice®HRD test (Myriad Genetics)<br>• GIS-score ≥42<br>• t*BRCA* mutations | All patients<br>t*BRCA* mt<br>GIS-high or t*BRCA* mt<br>GIS high and t*BRCA*wt<br>GIS-low and t*BRCA*wt | 22.1 vs. 16.6 : 0.59 (0.49–0.72)<br>37.2 vs. 21.7 : 0.31 (0.2–0.47)<br>37.2 vs. 17.7 : 0.33 (0.25–0.45)<br>28.1 vs. 16.6 : 0.43 (0.28–0.66)<br>16.9 vs. 16 : 0.92 (0.72–1.17) |

(continued)

**Table 1.1** (continued)

| | Patients | No. | Intervention | Stratification | Biomarkers | Population | Median PFS (months): HR (95% CI) |
|---|---|---|---|---|---|---|---|
| PRIMA | • Newly diagnosed ovarian, peritoneal, fallopian-tubal cancer<br>• Stage III disease with visible residual tumor after debulking surgery, inoperable stage III disease, or any stage IV disease<br>• High-grade serous or endometrioid histologies<br>• CR/PR to platinum-based chemotherapy | 733 | • Maintenance olaparib (487)<br>vs.<br>• Maintenance placebo (246)<br>Until PD (up to 36 months) | • Clinical response after platinum-based chemotherapy (CR or PR)<br>• Neoadjuvant chemotherapy (yes or no)<br>• HR status (deficient vs. proficient or not determined). | myChoice®HRD test (Myriad Genetics)<br>• GIS-score ≥42<br>• t*BRCA* mutations | All patients<br>t*BRCA* mt<br>GIS-high or t*BRCA* mt<br>GIS-high and t*BRCA*wt<br>GIS-low and t*BRCA*wt | 13.8 vs. 8.2 : 0.62 (0.50–0.76)<br>22.1 vs. 10.9: 0.4 (0.27-0.62)<br>21.9 vs. 10.4: 0.43 (0.31-0.59)<br>19.6 vs. 8.2: 0.5 (0.31-0.83)<br>8.1 vs. 5.1: 0.68 (0.49-0.94) |
| VELIA | • Newly diagnosed ovarian, peritoneal, fallopian-tubal cancer<br>• Stage III or IV<br>• High-grade serous histology | 1140 | • Carboplatine/taxane with placebo followed by maintenance placebo (control) (375)<br>vs.<br>• Carboplatine/taxane with veriparib followed by maintenance placebo (veliparib combination) (383)<br>vs.<br>• Carboplatine/taxane with veriparib followed by maintenance veriparib (velaparib throughout) (382)<br>Until PD (up to 36x3-weeks cycles of platinum-based chemotherapy and 30 cycles of maintenance therapy) | • Timing of surgery<br>• Residual tumor after primary surgery<br>• Paclitaxel schedule<br>• Geographic region<br>• Stage of disease<br>• Germline *BRCA* status | myChoice®HRD Plus assay (Myriad Genetics)<br>• GIS-score ≥ 33 (genomic scar)<br>• tBRCA mutations | All patients<br>t*BRCA* mt<br>GIS-high or t*BRCA*mt<br>t*BRCA* wt<br>GIS-low and t*BRCA*wt | (veliparib throughout vs. control)<br>23.5 vs. 17.3 : 0.68 (0.56–0.83)<br>34.7 vs. 22 : 0.44 (0.28–0.68)<br>31.9 vs. 20.5 : 0.57 (0.43–0.76)<br>18.2 vs. 15.1 : 0.80 (0.64–1.00)<br>15.0 vs. 11.5 : 0.81 (0.6–1.09) |

| | | | | | | |
|---|---|---|---|---|---|---|
| *Maintenance therapy in platinum-sensitive recurrence* | | | | | | |
| Study 19 | • Recurrent ovarian or fallopian tube cancer or primary peritoneal cancer<br>• Tumor with high-grade (grade 2 or 3) serous features or a serous component<br>• Platinum-sensitive (defined by an objective response to a previous platinum-based therapy for more than 6 months)<br>• Completion of at least two courses of platinum-based chemotherapy<br>• Objective response to their most recent regimen (CR/PR) | 265 | • Maintenance of olaparib after completion of the last dose of platinum-based chemotherapy (400 mg capsules, twice daily) (136)<br>vs.<br>• Maintenance placebo after completion of the last dose of platinum-based chemotherapy (129)<br>Until PD | • Length of platinum-free interval(>6-12months/>12months)<br>• Objective response to their most recent regimen (CR vs. PR)<br>• Ancestry to help the distribution of BRCA1/2 germline mutations | Myriad BRCAnalysis (Myriad Genetics)<br>tBRCA testing (Foundation Medicine) | All patients<br>t*BRCA* mt<br>t*BRCA* wt | 10.8 vs. 5.4 : 0.35 (0.25–0.49)<br>11.2 vs. 4.3 : 0.18 (0.1–0.31)<br>7.4 vs. 5.5 : 0.54 (0.34–0.85) |
| SOLO2 | • Recurrent high-grade serous or high-grade endometrioid ovarian, peritoneal or fallopian tube cancer<br>• Platinum-sensitive disease (disease progression occurring at least 6 months after the last dose of platinum therapy was given)<br>• Completion of at least two courses of platinum-based chemotherapy<br>• Objective response to their most recent regimen (CR/PR)<br>• Predicted or suspected deleterious, BRCA1/2 mutation | 295 | • Maintenance olaparib (300 mg in two 150 mg tablets, twice daily) (196)<br>vs.<br>• Maintenance placebo (99)<br>Until PD | • Response to previous chemotherapy (PR/CR)<br>• Length of platinum-free interval<br>(>6–12 months / >12 months) | Myriad BRCAnalysis (Myriad Genetics) | All patients<br>All patients | 19.1 vs. 5.5 : 0.33 (0.24-0.44)<br>OS<br>51.7 vs. 38.8 : 0.74 (0.54-1.00) |

(continued)

**Table 1.1** (continued)

| | Patients | No. | Intervention | Stratification | Biomarkers | Population | Median PFS (months): HR (95% CI) |
|---|---|---|---|---|---|---|---|
| NOVA | • Histologically diagnosed ovarian cancer, fallopian tube cancer, or primary peritoneal cancer<br>• High-grade serous histologic features<br>• Platinum-sensitive reccurrence<br>• Completion of at least two courses of platinum-based chemotherapy<br>• Objective response to their most recent regimen (CR/PR) | 553 | • Maintenance niraparib (300 mg tablets, daily) (372)<br>vs.<br>• Maintenance placebo (181)<br>Until PD | • Length of platinum-free interval (>6-12 months / >12 months)<br>• Use of bevacizumab in conjunction with the penultimate or last platinum regimen<br>• Response to previous chemotherapy (PR/CR) | Myriad BRCAnalysis (Myriad Genetics)<br>myChoice®HRD test (Myriad Genetics) | g*BRCA* mt<br>GIS-high and g*BRCA* wt<br>g*BRCA* wt<br>GIS-low and *BRCA* wt | 21 vs. 5.5 : 0.27 (0.17-0.41)<br>12.9 vs. 3.8 : 0.38 (0.24-0.59)<br>9.3 vs. 3.9 : 0.45 (0.34-0.61)<br>6.9 vs. 3.8 : 0.58 (0.34-0.61) |
| ARIEL3 | • Histologically diagnosed ovarian cancer, fallopian tube cancer, or primary peritoneal cancer<br>• High-grade serous and endometrioid histologies<br>• Platinum-sensitive reccurrence<br>• Completion of at least two courses of platinum-based chemotherapy<br>• Objective response to their most recent regimen (CR/PR) | 564 | • Maintenance rucaparib (600 mg twice, daily) (375)<br>vs.<br>• Maintenance placebo (189)<br>Until PD | • HRR gene mutation status<br>• Length of platinum-free interval (>6-12 months / >12 months)<br>• Response to previous chemotherapy (PR/CR) | T5 NGS assay (Foundation Medicine)<br>• Genomic LOH<br>• tBRCA mt<br>• HRR pathway gene mt<br>Myriad BRCAnalysis (Myriad Genetics) | All patients<br>g/t*BRCA* mt<br>LOH-high or g/t*BRCA* mt<br>LOH-high and *BRCA* wt<br>LOH-low and *BRCA* wt | 10.8 vs. 5.4 : 0.37 (0.3-0.45)<br>16.6 vs. 5.4 : 0.23 (0.16-0.34)<br>13.6 vs. 5.4 : 0.32 (0.24-0.42)<br>9.7 vs. 5.4 : 0.44 (0.29-0.66)<br>6.7 vs. 5.4 : 0.58 (0.4-0.8) |

*NED* no evidence of disease, *CR* complete response, *PR* partial response, *PD* progressive disease, *PFS* progression-free survival, *OS* overall survival, *GIS* genomic instability score, *wt* wild type, *mt* mutation, *HRD* homologous recombination deficiency, *HR* homologous recombination repair, *tBRCA* tumor *BRCA*, *gBRCA* germline *BRCA*, *LOH* loss of heterozygosity clinical risk (higher-risk: stage IV, stage III with residual tumor at primary debulking surgery (PDS) or disease performed interval debulking surgery (IDS), lower-risk: stage III disease with no residual tumor after PDS), significant improvement of PFS was observed in both groups of higher-risk (median PFS: 40.6 months vs. 11.1 months; HR: 0.34; 95% CI: 0.24–0.49) and lower-risk (median PFS: not reached vs. 21.9 months; HR: 0.38; 95% CI: 0.25–0.59) [20]. Moreover, other detailed subgroup analyses stratified by surgical timing, residual tumor, response after platinum-based chemotherapy, *BRCA* mutational status, also confirmed the survival benefits of olaparib maintenance therapy in patients with g*BRCA* mutation [21]. PAOLA-1 was a phase III randomized control study investigating the efficacy of combined maintenance therapy with Bev and olaparib after the response to first-line platinum-based chemotherapy in patients with stages III and IV disease of HGSOC and EMOC [22]. Significant improvement of PFS was observed in patients who received the combined maintenance therapy with Bev and olaparib compared with those who received single agent maintenance with Bev (median PFS: 22.1 months vs. 16.6 months; HR: 0.59; 95% CI: 0.49–0.72). In the subgroup analysis, a significant improvement of PFS was observed in patients with tumor *BRCA* mutation (t*BRCA*) (HR: 0.31; 95% CI: 0.20–0.47) and those with HRD tumors including t*BRCA* mutation (HR: 0.33; 95% CI: 0.25–0.45). In contrast, no survival benefit was observed in the subgroup with HR proficient (HRP) compared with placebo group (median PFS: 16.6 months vs. 16.2 months; HR: 1.00; 95% CI: 0.75–1.35). Additionally, subgroup analyses stratified by higher-risk, lower-risk, higher-risk with t*BRCA* mutation, lower-risk with t*BRCA* mutation, higher-risk with HRD and lower-risk with HRD subgroups showed improvement of PFS. In particular, the PFS rate at 24 months was 95.5% and 89.7% in lower-risk with t*BRCA* mutation and lower-risk with HRD, respectively [23]. The PRIMA study evaluated the efficacy of niraparib maintenance therapy following first-line adjuvant chemotherapy in stages III and IV disease after a response to platinum-based chemotherapy [24]. The important point of the population in the PRIMA study was that only patients with stage III disease with visible residual tumor after PDS, inoperable stage III disease, or any stage IV disease, were recruited. In the overall population, significant benefit of PFS was observed in niraparib maintenance group compared with placebo group (median PFS: 13.8 months vs. 8.2 months; HR: 0.62; 95% CI: 0.50–0.76). Moreover, improvement of PFS was seen in not only patients with HRD (median PFS: 21.9 months vs. 10.4 months; HR: 0.43; 95% CI: 0.31–0.59) but also in those with HRP tumors (median PFS: 8.1 months vs. 5.4 months; HR: 0.68; 95% CI: 0.49–0.94). The clinical benefit of first-line combination and maintenance therapy with veliparib (veliparib throughout) was assessed in the VELIA study in patients with previously untreated stages III and IV HGSOC [25]. PFS was statistically improved in patients with veliparib throughout patients compared with those receiving combination and maintenance placebo (control) (median PFS: 23.5 months vs. 17.3 months; HR: 0.68; 95% CI: 0.56–0.83). In the subgroup analysis, statistically significant benefit of PFS was observed in the patients with g*BRCA* mutation (median PFS: 34.7 months vs. 22.0 months; HR: 0.44; 95% CI: 0.28–0.68) or those with HRD (median PFS: 31.9 months vs. 20.5 months; HR: 0.57; 95% CI: 0.43–0.76), compared with control group. However, no survival benefit of veliparib maintenance therapy was observed in subgroup with *BRCA* wild-type (*BRCA*wt) (median PFS: 18.2 months vs. 15.1 months; HR: 0.80; 95% CI: 0.64–1.00) or patients with HRP (median PFS: 15.0 months vs. 1.5 months; HR: 0.81; 95% CI: 0.60–1.09). Veliparib use has not been approved in the first-line setting at the moment. Based on the results of these phase III clinical trials, ASCO guidelines of PARP inhibitors in the management of OC recommended olaparib (for those with pathogenic or likely pathogenic variants in *BRCA1/2* genes) or niraparib (for all patients) maintenance therapy in patients with newly diagnosed stage III-IV OC who showed CR/PR to first-line platinum-based chemotherapy, and combined maintenance therapy with olaparib and Bev in those with HRD [26]. However, it has been unclear whether the addition of combined and maintenance Bev could increase clinical benefit compared with olaparib maintenance therapy following platinum-based chemotherapy, because the arm of platinum-based chemotherapy followed by olaparib maintenance was not included in the PAOLA-1 study. To address this question, a population-adjusted indirect comparison was performed, and increased benefit with Bev addition was shown in newly diagnosed advanced OC with *BRCA* mutation [27]. However, indirect comparison is limited, and a further well-designed clinical trial should be performed to answer this question. Another key question is how PARP inhibitors should be used in treatment for newly diagnosed advanced OC with HRP. HRP patients were included in participants of PAOLA-1, PRIMA, and VELIA trials, and significant improvement of PFS in this population was observed in only PRIMA, but not in other two studies. One of the reasons for the effectiveness of niraparib for not only HRD but also HRP may be due to the difference in eligibility criteria. Since the PAOLA-1 study enrolled patients regardless of presence of residual tumor after PDS and surgical timing (PDS or IDS), Clinically higher-risk patients, such as stage III disease with visible residual tumor after PDS, inoperable stage III disease, or any stage IV disease, were subjected to the PRIMA study. CR rate of the patients was 20% and 69% in PAOLA-1 and PRIMA, respectively, although 53% of NED was included in the PAOLA-1 population. Therefore, PRIMA population might have deeper platinum-sensitivity compared with PAOLA-1 population. Another reason is the molecular or pharmacokinetic difference between niraparib and other PARP inhibitors. Niraparib is characterized by higher tumor penetration and higher PARP trapping ability compared with olaparib and rucaparib [28]. And recent basic studies showed tumor exposure to niraparib is 3.3 times greater than plasma exposure in tumor xenograft mouse models [29].

PARP inhibitors in second-line and beyond maintenance settings have assessed the benefit of survival in four clinical trials (Table 1.1). Study19 investigated the efficacy of olaparib maintenance therapy in patients with platinum-sensitive, recurrent, high-grade OC who had received ≥2 prior lines of platinum-based chemotherapy and had a CR/PR to the most recent treatment [30]. Olaparib maintenance therapy demonstrated clinical benefit of PFS compared with placebo maintenance arm (median PFS: 8.4 months vs. 4.8 months; HR: 0.35; 95% CI: 0.25–0.49). *BRCA* mutation was detected in 51% of participants in the study. PFS improvement by olaparib maintenance therapy was also observed in the patients with g*BRCA* mutation compared with those who received placebo maintenance (median PFS: 11.2 months vs. 4.3 months; HR: 0.18; 95% CI: 0.1–0.31). In the analysis for OS after a median follow-up of 78 months, slight survival benefit for PFS was observed in both groups of all patients (median OS: 29.8 months vs. 27.8 months; HR: 0.73; 95% CI: 0.55–0.96) and those with g*BRCA* mutation (median OS: 34.9 months vs. 30.2 months; HR: 0.62; 95% CI: 0.41–0.94), although the p-value was 0.025, the threshold was not set to determine the statistical significance for the analysis, suggesting the p value is nominal [31]. SOLO2 trial demonstrated the efficacy of maintenance therapy in platinum-sensitive recurrent HGSOC and EMOC patients with completion of at least two courses of platinum-based chemotherapy, CR/PR to most recent platinum-based chemotherapy and deleterious g*BRCA* mutation [32]. PFS showed significant improvement for the olaparib group compared with placebo group (median PFS: 19.1 months vs. 5.5 months; HR, 0.30; 95% CI, 0.22 to 0.41). In the final overall survival (OS) analysis in the full analysis set, the median OS was 51.7 and 38.8 months in olaparib and placebo patients, respectively (HR, 0.74; 95% CI, 0.54 to 1.00) [16]. Additionally, in patients with a g*BRCA* mutation confirmed by Myriad Genetics BRCA test, median OS was 52.4 months and 37.4 months in olaparib and placebo (HR, 0.71; 95% CI, 0.52 to 0.97). NOVA trial demonstrated clinical benefit of maintenance niraparib in patients with platinum-sensitive recurrent OC who had a CR/PR after ≥2 prior regimens with platinum-based chemotherapy [33]. Only HGSOC was enrolled in the study. PFS was significantly improved in niraparib compared with placebo in patients with g*BRCA* mutation (median PFS: 21.0 months vs. 5.5 months; HR: 0.27; 95% CI: 0.17–0.41), HRD with *BRCA*wt (median PFS: 12.9 months vs. 3.8 months; HR: 0.38; 95% CI: 0.24–0.59), non-g*BRCA* mutation (median PFS: 9.3 months vs. 3.9 months; HR, 0.45: 95% CI: 0.34–0.61), and HRP (median PFS: 6.9 months vs. 3.8 months; HR: 0.58; 95% CI: 0.36–0.92). Similarly, the efficacy of rucaparib was confirmed in ARIEL3 trial in patients with platinum-sensitive, HGSOC or EMOC histologies and recurrent OC who had a CR/PR after ≥2 prior regimens with platinum-based chemotherapy [34]. Improvement of PFS were confirmed in overall patients (median PFS: 10.8 months vs. 5.4 months; HR: 0.37; 95% CI: 0.3–0.45), g/t*BRCA* mutation (median PFS: 16.6 months vs. 5.4 months; HR, 0.23: 95% CI: 0.16–0.34), LOH-high or g/t*BRCA* mutation (median PFS: 13.6 months vs. 5.4 months; HR: 0.32; 95% CI: 0.24–0.42), LOH-high and *BRCA*wt (median PFS: 9.7 months vs. 5.4 months; HR, 0.44; 95% CI, 0.29 to 0.66) and LOH-low and *BRCA*wt (median PFS: 6.7 months vs. 5.4 months; HR: 0.58; 95% CI: 0.4–0.8). Based on the results of these clinical trials, PARP inhibitor monotherapy maintenance using olaparib, niraparib, or rucaparib was recommended for treatment in patients with OC who have responded to platinum-based therapy regardless of *BRCA* mutation status and who have a pathogenic or likely pathogenic g/t*BRCA* mutation [26].

Monotherapy treatment with PARP inhibitors may be a possible treatment for relapsed patients with platinum-sensitivity and HRD confirmed by Myriad myChoice CDx, due to several clinical trials with olaparib having shown the improvement of PFS. In SOLO3 trial, the objective response (ORR) of olaparib and physician's choice chemotherapy was 72% and 51%, respectively (odds ratio, 2.53;

95% CI: 1.40–4.58; $P = 0.002$) in patients with g*BRCA* mutation [35]. Moreover, improvement of PFS was shown in olaparib compared with physician's choice chemotherapy in analysis according to both independent central review (median PFS: 13.4 months vs. 9.2 months; HR: 0.62; 95% CI: 0.43–0.91) and investigator assessment (median PFS: 13.2 months vs. 8.5 months; HR: 0.49; 95% CI: 0.35–0.70). In terms of niraparib, a single-arm nonrandomized trial, QUADRA trial, studied patients with relapsed HGSOC who had been treated with ≥3 prior chemotherapies. ORR was 28% (95% CI: 15.6% to 42.6%) in those with HRD disease who received three or four previous chemotherapy and were sensitive to most recent platinum-based chemotherapy [36]. In addition, the median PFS was 5.5 months (95% CI: 3.5 to 8.2 months), median duration of response was 9.2 months (95% CI: 5.9 to not estimable months) and median OS was 17.2 months (95% CI: 14.9 to 19.8 months). Similarly, the efficacy of rucaparib monotherapy was assessed in two single-arm clinical trials, such as ARIEL2 [37] and Study10 [38], was confirmed in measurable disease with g*BRCA* mutation.

Considering the results of the clinical trials described above, the efficacy of PARP inhibitors seemed to be broadly accepted in the setting of first-line and second or more lines in OC patients with platinum-sensitive, g/t*BRCA* mutation and HRD status. However, treatment of any PARP inhibitor in patients with *BRCA*wt, HRP status and platinum-resistant recurrent OC is not currently recommended. Moreover, a significant limitation of clinical trials of PARPi is that the efficacy of maintenance therapy has not been accurately investigated in CCC and MC. Most clinical trials have focused on patients with HGSOC and/or EMOC, and did not include patients with histologies of CCC and MC. Although, it has been reported that the frequency of HRD including *BRCA* mutation has been rare, several studies have shown a small population of CCC and MC showed molecular alteration of HRD including *BRCA* mutation, suggesting that maintenance therapy for these patients could be a treatment option for a small subset of CCC and MC patients.

## 1.4 Immune Checkpoint Inhibitors

Immune checkpoint inhibitors (ICIs) can be effective against OC and are expected to become a new treatment option. Tumor-infiltrating T cells have been reported to be an independent prognostic factor in OC [39], however, the response rate has only been about 10–15%. CCC may be highly sensitive to ICIs. The therapeutic effects of ICIs for OC include anti-programmed death-1 (anti-PD-1) antibodies (nivolumab and pembrolizumab) and anti-programmed death-ligand 1 (anti-PD-L1) antibodies (avelumab, durvalumab, and atezolizumab) are summarized.

A single-center, phase II trial of nivolumab in 20 platinum-resistant OC patients reported that the best overall response was 15%, which included 2 cases of complete response. The disease control rate (DCR) was 45% [40]. A subsequent phase III trial (NINJA) comparing nivolumab with second-line chemotherapy in 300 platinum-resistant OC patients suggested that nivolumab did not improve OS and showed worse PFS [41]. A phase III trial (ATHENA-COMBO) comparing the combination of rucaparib plus nivolumab with rucaparib alone following frontline platinum-based chemotherapy is being conducted [42]. Pembrolizumab is approved for any unresectable or metastatic solid tumor with microsatellite instability (MSI). A phase II trial (KEYNOTE-100) of pembrolizumab in 376 platinum-resistant OC patients reported that the ORR were 17% and 15.8% in the tumor PD-L1 expression (CPS) and CCC groups, although the overall ORR was only 8% [43]. A phase I/II trial of pembrolizumab plus niraparib in 60 platinum-resistant OC patients (TOPACIO/KEYNOTE-162) showed that the response rate was 25%, DCR 67%, and median duration of response (DOR) was 9.3 months [44]. Efficacy did not correlate with t*BRCA* mutation, HRD, or PD-L1 expression. A phase III trial (ENGOT-ov43) of pembrolizumab in combination with standard chemotherapy and pembrolizumab plus olaparib for maintenance therapy is underway in 1086 patients with primary advanced OC [45].

Avelumab was studied in combination with PLD in 900 platinum-resistant OC patients (JAVELIN Ovarian 200) [46] and in combination with chemotherapy/maintenance in 900 patients with previously untreated OC patients (JAVELIN Ovarian 100) [47]. However, no survival benefit was seen in either of these trials, which were suspended early. Durvalumab was shown to be effective in the treatment of platinum-sensitive recurrence with g*BRCA* mutation, with a response rate of 72% in combination with durvalumab and olaparib (MEDIOLA) [48]. A phase III trial of durvalumab in combination with chemotherapy and Bev followed by maintenance with Bev and olaparib in advanced OC patients (DUO-O) is ongoing [49]. A phase III study (IMaGYN050) of atezolizumab was conducted to evaluate the efficacy of atezolizumab in combination/maintenance with first-line chemotherapy plus Bev in 1300 newly diagnosed advanced OC patients, but it did not significantly prolong PFS [50].

At present, the efficacy of ICIs in OC has not been sufficiently demonstrated. In the future, it is hoped that biomarkers will be developed to identify patients with OC that respond to ICIs.

## 1.5 Antibody–Drug Conjugates

Antibody–drug conjugates (ADCs) which consist of an antibody, a payload, and a linker, have a strong affinity for the target tumor antigen [51, 52]. ADCs bind to a small molecule with cytotoxic drug via a linker and the compound can be released at the target site [53]. In other words, the ADC is like the train, the payload is the passenger, and the antigen is the station. A passenger descending at a station attacks cancer cells in a specific area. ADCs have been already used clinically in breast and hematological cancer [54, 55]. Several ADCs have reached clinical studies and are being evaluated in OC (Table 1.2). Target antigens are tumor antigens selected to be preferentially expressed on the membrane surface of tumor cells. Binding of the antibody to the antigen leads to internalization of the complex through endocytosis and lysosomal degradation, delivering the cytotoxic payload to the tumor cells.

**Table 1.2** List of clinical trials using ADCs

| Target antigen | Function | Expression in EOC | Related histology | ADCs | Payload action | Payloads | DAR | Linker | Phase | Including EOC patients |
|---|---|---|---|---|---|---|---|---|---|---|
| FRα | Intracellular transport of folate | 67–100% | HGSOC | Mirvetuximab Soravtansine | Microtubule inhibitor | DM-4 | 3–4 | Cleavable | III | Yes |
| | | | | STRO-002 | | Hemiasterlin | 4 | | I | No |
| | | | | MORAb-202 | | Eribulin | 4 | | I | Yes |
| Mesothelin | Cell adhesion | 55–100% | HGSOC, EMOC | Anetumab Ravtansine | | DM4 | 3.2 | | Ib | Yes |
| | | | | DMOT4039A | | MMAE | 3.5 | | I | Yes |
| | | | | BMS-986148 | | Tubulysin | 1.4 | | I/IIa | No |
| NaPi2b | Sodium-dependent surface transporter | 80–93% | HGSOC, EMOC | Upifitamab rilsodotin(XMT-1536) | | MMAF | 12–15 | | I | Yes |
| | | | | Lifastuzumab vedotin | | MMAE | 3–4 | | II | Yes |
| MUC16 | Protection of epithelial surfaces | 70–90% | – | Sofituzumab vedotin | | | 3.5 | | I | No |
| | | | | DMUC4064A | | | 2 | | I | Yes |
| Tissue factor | Coagulation cascade | 23–100% | – | Tisotumab vedotin | | | NR | | II | Yes |
| TIM1 | Immune responses | 90% | CCC | CDX-014 | | | 4.5 | | I | Yes |
| NOTCH-3 | Cell growth | 63% | HGSOC | PF-06650808 | | | NR | | I | No |
| AXL | Cell growth, invasion, metastasis | High frequency | – | Enapotamab vedotin | | | NR | | I-II | Yes |
| ALCAM/CD166 | Cell adhesion | Unknown | – | Praluzatamab ravtansine | | DM4 | 3.5 | | I | Yes |
| PKT7 | Polarity and adhesion | 28–69% loss | HGSOC | Cofetuzumab pelidotin | | Auristatin-0101 | NR | | I | No |
| DPEP3 | Membrane-bound | 25% | HGSOC | Tamrintamab pamozirin | DNA cross-linking | PBD | NR | | I | Yes |
| CLDN6 and 9 | Cell adhesion | 69% | HGSOC | SC-004 | | | NR | NR | I | Yes |
| TROP2 | Proliferation and invasion | 96% | EMOC | Sacituzumab govitecan | Topoisomerase inhibitor | SN-38 | 6.78 | Cleavable | I-II | No |

*FRα* Folate receptor alpha, *NaPi2b* Mesothelin, the sodium-dependent phosphate transport protein 2B, *MUC16* Mucin 16, *TIM1* Tissue Factor, T cell immunoglobulin and mucin domain-1, *NOTCH-3* notch homolog protein 3, *ALCAM/CD166* AXL, Activated Leukocyte Cell Adhesion Molecule, *PTK7* Protein Tyrosine Kinase 7, *CLDN6 and CLDN9* Claudin 6 and 9

Thus, choosing the correct antigen is the first step in making it effective [56]. The target antigens in OC include Folate receptor alpha (FRα), Mesothelin, the sodium-dependent phosphate transport protein 2B (NaPi2b), Mucin 16 (MUC16), Tissue Factor, T cell immunoglobulin and mucin domain-1 (TIM1), notch homolog protein 3 (NOTCH-3), AXL, Activated Leukocyte Cell Adhesion Molecule (ALCAM/CD166), Protein Tyrosine Kinase 7 (PTK7), Claudin 6 and 9 (CLDN6 and CLDN9), Tumor-associated calcium signal transducer 2 (TROP2), and Dipeptidase 3 (DPEP3). The functions of target antigens are diverse, including transporter, cell adhesion, cell growth, proliferation, invasion, and immune response. Target antigens with low expression in normal cells and high expression in cancer cells are thought to enhance selectivity, and many antigens are highly expressed in cancer cells. NaPi2b is frequently expressed in well-differentiated EMOC in addition to HGSOC [57]. TROP2 and TIM1 are frequently expressed in EMOC and CCC, respectively [58, 59]. Since only 1% of the administered ADC reaches the target tumor site, the ideal payload should be a small molecule with potent activity, and to date few molecules have been identified as optimal payload candidates for the conjugation process [60]. Functions of payloads in ADCs are classified as Microtubule inhibitor, DNA minor grove cross-linking, and Topoisomerase inhibitor. In addition, payloads include Monomethyl Auristatin E (MMAE), DM-4 (Ravtansine/soravtansine), Monomethyl Auristatin F (MMAF), Hemistarlin, Eribulin, Tubulysin, and Auristatin-0101. Payload delivery to targeted cells are limited by antigen expression and the average of drug molecules conjugated to the antibody, so-called drug-to-antibody ratio (DAR). In ADCs, linkers link antibodies to drugs, and ADCs must be kept stable because the bound drugs, which are very potent in serum, can cause side effects if released. Linkers can be classified as cleavable or non-cleavable. These linkers are popular in the ADC clinical pipeline with acid-sensitive linkers such as hydrazones and silyl ethers at the forefront. These linkers are chemically labile structures that can be cleaved depending on certain intracellular circumstances such as acid pH levels, high levels of glutathione, some cleavable linkers can also deliver the drug extracellularly, in the acid pH tumoral microenvironment, inducing killing in nearby tumor cells with no expression of the targeted antigen. Most of the ADC linkers associated with OC are cleavable linkers. Although majority are Phase I clinical trials, the clinical trial for Mirvetuximab soravtansine was in Phase III [61, 62]. This phase III study carried out on 366 patients with OC demonstrated a clinical benefit of Mirvetuximab soravtansine compared to chemotherapy. In terms of efficacy, although a significant improvement in PFS neither Mirvetuximab soravtansine nor chemotherapy-treated patients were appreciated, and responses rates were higher in the patients treated with the ADC. In addition, CDX-014, which targets TIM1, which is highly expressed in CCC, has completed PI and is awaiting further development. In addition, Sacituzumab govitecan, which targets TROP2 and a Topoisomerase inhibitor payload, which is frequently expressed in endometrial cancer, has been tested in endometrial cancer and may be expected to be applied to EMOC.

Although ADCs in OC are not fully developed for clinical use yet, their development is progressing. Due to their unique mechanisms of action, ADCs have the potential to provide precision therapeutic benefit to populations expressing appropriate target antigens for which angiogenesis inhibitors, PARP inhibitors, and ICIs have failed to improve prognosis.

## 1.6 Conclusion

New therapies incorporating single-agent or combination therapy with the molecular targeted agents reviewed in this chapter will prove beneficial to a significant proportion of OC patients. In the same way that current practice guidelines are based on evidence from several phase III clinical trials, further clinical trials focusing on these molecular targeted agents may contribute to the establishment of a new standard of care for patients with OC.

## References

1. Burger RA, Brady MF, Bookman MA, Fleming GF, Monk BJ, Huang H, et al. Incorporation of bevacizumab in the primary treatment of ovarian cancer. N Engl J Med. 2011;365:2473–83.
2. Perren TJ, Swart AM, Pfisterer J, Ledermann JA, Pujade-Lauraine E, Kristensen G, et al. A phase 3 trial of bevacizumab in ovarian cancer. N Engl J Med. 2011;365:2484–96.
3. Tewari KS, Burger RA, Enserro D, Norquist BM, Swisher EM, Brady MF, et al. Final overall survival of a randomized trial of bevacizumab for primary treatment of ovarian cancer. J Clin Oncol. 2019;37:2317–28.
4. Oza AM, Cook AD, Pfisterer J, Embleton A, Ledermann JA, Pujade-Lauraine E, et al. Standard chemotherapy with or without bevacizumab for women with newly diagnosed ovarian cancer (ICON7): overall survival results of a phase 3 randomized trial. Lancet Oncol. 2015;16:928–36.
5. Aghajanian C, Blank SV, Goff BA, Judson PL, Teneriello MG, Husain A, et al. OCEANS: a randomized, double-blind, placebo-controlled phase III trial of chemotherapy with or without bevacizumab in patients with platinum-sensitive recurrent epithelial ovarian, primary peritoneal, or fallopian tube cancer. J Clin Oncol. 2012;30:2039–45.
6. Aghajanian C, Goff B, Nycum LR, Wang YV, Husain A, Blank SV. Final overall survival and safety analysis of OCEANS, a phase 3 trial of chemotherapy with or without bevacizumab in patients with platinum-sensitive recurrent ovarian cancer. Gynecol Oncol. 2015;139:10–6.
7. Coleman RL, Brady MF, Herzog TJ, Sabbatini P, Armstrong DK, Walker JL, et al. Bevacizumab and paclitaxel-carboplatin chemotherapy and secondary cytoreduction in recurrent, platinum-sensitive ovarian cancer (NRG oncology/gynecologic oncology group study GOG-0213): a multicentre, open-label, randomised, phase 3 trial. Lancet Oncol. 2017;18:779–91.
8. Pfisterer J, Shannon CM, Baumann K, Rau J, Harter P, Joly F, et al. Bevacizumab and platinum-based combinations for recurrent ovarian cancer: a randomised, open-label, phase 3 trial. Lancet Oncol. 2020;21:699–709.
9. Pignata S, Lorusso D, Joly F, Gallo C, Colombo N, Sessa C, et al. Carboplatin-based doublet plus bevacizumab beyond progression versus carboplatin-based doublet alone in patients with platinum-sensitive ovarian cancer: a randomised, phase 3 trial. Lancet Oncol. 2021;22:267–76.
10. Pujade-Lauraine E, Hilpert F, Weber B, Reuss A, Poveda A, Kristensen G, et al. Bevacizumab combined with chemotherapy for platinum-resistant recurrent ovarian cancer: the AURELIA open-label randomized phase III trial. J Clin Oncol. 2014;32:1302–8.

11. Bais C, Mueller B, Brady MF, Mannel RS, Burger RA, Wei W, et al. Tumor microvessel density as a potential predictive marker for bevacizumab benefit: GOG-0218 biomarker analyses. J Natl Cancer Inst. 2017;109:djx066.
12. Alvarez Secord A, Bell Burdett K, Owzar K, Tritchler D, Sibley AB, Liu Y, et al. Predictive blood-based biomarkers in patients with epithelial ovarian cancer treated with carboplatin and paclitaxel with or without bevacizumab: results from GOG-0218. Clin Cancer Res. 2020;26:1288–96.
13. Collinson F, Hutchinson M, Craven RA, Cairns DA, Zougman A, Wind TC, et al. Predicting response to bevacizumab in ovarian cancer: a panel of potential biomarkers informing treatment selection. Clin Cancer Res. 2013;19:5227–39.
14. Norquist BM, Brady MF, Harrell MI, Walsh T, Lee MK, Gulsuner S, et al. Mutations in homologous recombination genes and outcomes in ovarian carcinoma patients in GOG 218: an NRG oncology/gynecologic oncology group study. Clin Cancer Res. 2018;24:777–83.
15. Ferriss JS, Java JJ, Bookman MA, Fleming GF, Monk BJ, Walker JL, et al. Ascites predicts treatment benefit of bevacizumab in front-line therapy of advanced epithelial ovarian, fallopian tube and peritoneal cancers: an NRG oncology/GOG study. Gynecol Oncol. 2015;139:17–22.
16. Wade KNS, Brady MF, Thai T, Wang Y, Zheng B, Salani R, et al. Measurements of adiposity as prognostic biomarkers for survival with anti-angiogenic treatment in epithelial ovarian cancer: an NRG oncology/gynecologic oncology group ancillary data analysis of GOG 218. Gynecol Oncol. 2019;155:69–74.
17. Buechel ME, Enserro D, Burger RA, Brady MF, Wade K, Secord AA, et al. Correlation of imaging and plasma based biomarkers to predict response to bevacizumab in epithelial ovarian cancer (OC). Gynecol Oncol. 2021;161:382–8.
18. O'Connor MJ. Targeting the DNA damage response in cancer. Mol Cell. 2015;60:547–60.
19. Moore K, Colombo N, Scambia G, Kim BG, Oaknin A, Friedlander M, et al. Maintenance Olaparib in patients with newly diagnosed advanced ovarian cancer. N Engl J Med. 2018;379:2495–505.
20. Banerjee S, Moore KN, Colombo N, Scambia G, Kim BG, Oaknin A, et al. Maintenance olaparib for patients with newly diagnosed advanced ovarian cancer and a BRCA mutation (SOLO1/GOG 3004): 5-year follow-up of a randomised, double-blind, placebo-controlled, phase 3 trial. Lancet Oncol. 2021;22:1721–31.
21. DiSilvestro P, Colombo N, Scambia G, Kim BG, Oaknin A, Friedlander M, et al. Efficacy of maintenance Olaparib for patients with newly diagnosed advanced ovarian cancer with a BRCA mutation: subgroup analysis findings from the SOLO1 trial. J Clin Oncol. 2020;38:3528–37.
22. Ray-Coquard I, Pautier P, Pignata S, Perol D, Gonzalez-Martin A, Berger R, et al. Olaparib plus bevacizumab as first-line maintenance in ovarian cancer. N Engl J Med. 2019;381:2416–28.
23. Harter P, Mouret-Reynier MA, Pignata S, Cropet C, Gonzalez-Martin A, Bogner G, et al. Efficacy of maintenance olaparib plus bevacizumab according to clinical risk in patients with newly diagnosed, advanced ovarian cancer in the phase III PAOLA-1/ENGOT-ov25 trial. Gynecol Oncol. 2022;164:254–64.
24. Gonzalez-Martin A, Pothuri B, Vergote I, DePont CR, Graybill W, Mirza MR, et al. Niraparib in patients with newly diagnosed advanced ovarian cancer. N Engl J Med. 2019;381:2391–402.
25. Coleman RL, Fleming GF, Brady MF, Swisher EM, Steffensen KD, Friedlander M, et al. Veliparib with first-line chemotherapy and as maintenance therapy in ovarian cancer. N Engl J Med. 2019;381:2403–15.
26. Tew WP, Lacchetti C, Ellis A, Maxian K, Banerjee S, Bookman M, et al. PARP inhibitors in the Management of Ovarian Cancer: ASCO guideline. J Clin Oncol. 2020;38:3468–93.
27. Vergote I, Ray-Coquard I, Anderson DM, Cantuaria G, Colombo N, Garnier-Tixidre C, et al. Population-adjusted indirect treatment comparison of the SOLO1 and PAOLA-1/ENGOT-ov25 trials evaluating maintenance olaparib or bevacizumab or the combination of both in newly diagnosed, advanced BRCA-mutated ovarian cancer. Eur J Cancer. 2021;157:415–23.
28. Murai J, Huang S-YN, Renaud A, Zhang Y, Ji J, Takeda S, et al. Stereospecific PARP trapping by BMN 673 and comparison with Olaparib and Rucaparib. Mol Cancer Ther. 2014;13:433–43.

29. Sun K, Mikule K, Wang Z, Poon G, Vaidyanathan A, Smith G, et al. A comparative pharmacokinetic study of PARP inhibitors demonstrates favorable properties for niraparib efficacy in preclinical tumor models. Oncotarget. 2018;9:37080–96.
30. Ledermann J, Harter P, Gourley C, Friedlander M, Vergote I, Rustin G, et al. Olaparib maintenance therapy in platinum-sensitive relapsed ovarian cancer. N Engl J Med. 2012;366:1382–92.
31. Ledermann JA, Harter P, Gourley C, Friedlander M, Vergote I, Rustin G, et al. Overall survival in patients with platinum-sensitive recurrent serous ovarian cancer receiving olaparib maintenance monotherapy: an updated analysis from a randomised, placebo-controlled, double-blind, phase 2 trial. Lancet Oncol. 2016;17:1579–89.
32. Pujade-Lauraine E, Ledermann JA, Selle F, Gebski V, Penson RT, Oza AM, et al. Olaparib tablets as maintenance therapy in patients with platinum-sensitive, relapsed ovarian cancer and a BRCA1/2 mutation (SOLO2/ENGOT-Ov21): a double-blind, randomised, placebo-controlled, phase 3 trial. Lancet Oncol. 2017;18:1274–84.
33. Mirza MR, Monk BJ, Herrstedt J, Oza AM, Mahner S, Redondo A, et al. Niraparib maintenance therapy in platinum-sensitive, recurrent ovarian cancer. N Engl J Med. 2016;375:2154–64.
34. Coleman RL, Oza AM, Lorusso D, Aghajanian C, Oaknin A, Dean A, et al. Rucaparib maintenance treatment for recurrent ovarian carcinoma after response to platinum therapy (ARIEL3): a randomised, double-blind, placebo-controlled, phase 3 trial. Lancet. 2017;390:1949–61.
35. Penson RT, Valencia RV, Cibula D, Colombo N, Leath CA 3rd, Bidzinski M, et al. Olaparib versus nonplatinum chemotherapy in patients with platinum-sensitive relapsed ovarian cancer and a germline BRCA1/2 mutation (SOLO3): a randomized phase III trial. J Clin Oncol. 2020;38:1164–74.
36. Moore KN, Secord AA, Geller MA, Miller DS, Cloven N, Fleming GF, et al. Niraparib monotherapy for late-line treatment of ovarian cancer (QUADRA): a multicentre, open-label, single-arm, phase 2 trial. Lancet Oncol. 2019;20:636–48.
37. Swisher EM, Lin KK, Oza AM, Scott CL, Giordano H, Sun J, et al. Rucaparib in relapsed, platinum-sensitive high-grade ovarian carcinoma (ARIEL2 part 1): an international, multicentre, open-label, phase 2 trial. Lancet Oncol. 2017;18:75–87.
38. Kristeleit R, Shapiro GI, Burris HA, Oza AM, LoRusso P, Patel MR, et al. A phase I-II study of the Oral PARP inhibitor Rucaparib in patients with germline BRCA1/2-mutated ovarian carcinoma or other solid tumors. Clin Cancer Res. 2017;23:4095–106.
39. Zhang L, Conejo-Garcia JR, Katsaros D, Gimotty PA, Massobrio M, Regnani G, et al. Intratumoral T cells, recurrence, and survival in epithelial ovarian cancer. N Engl J Med. 2003;348:203–13.
40. Hamanishi J, Mandai M, Ikeda T, Minami M, Kawaguchi A, Murayama T, et al. Safety and antitumor activity of anti-PD-1 antibody, Nivolumab, in patients with platinum-resistant ovarian cancer. J Clin Oncol. 2015;33:4015–22.
41. Hamanishi J, Takeshima N, Katsumata N, Ushijima K, Kimura T, Takeuchi S, et al. Nivolumab versus gemcitabine or Pegylated liposomal doxorubicin for patients with platinum-resistant ovarian cancer: open-label, randomized trial in Japan (NINJA). J Clin Oncol. 2021;39:3671–81.
42. Monk BJ, Coleman RL, Fujiwara K, Wilson MK, Oza AM, Oaknin A, et al. ATHENA (GOG-3020/ENGOT-ov45): a randomized, phase III trial to evaluate rucaparib as monotherapy (ATHENA-MONO) and rucaparib in combination with nivolumab (ATHENA-COMBO) as maintenance treatment following frontline platinum-based chemotherapy in ovarian cancer. Int J Gynecol Cancer. 2021;31:1589–94.
43. Matulonis UA, Shapira-Frommer R, Santin AD, Lisyanskaya AS, Pignata S, Vergote I, et al. Antitumor activity and safety of pembrolizumab in patients with advanced recurrent ovarian cancer: results from the phase II KEYNOTE-100 study. Ann Oncol. 2019;30:1080–7.
44. Konstantinopoulos PA, Waggoner S, Vidal GA, Mita M, Moroney JW, Holloway R, et al. Single-arm phases 1 and 2 trial of Niraparib in combination with Pembrolizumab in patients with recurrent platinum-resistant ovarian carcinoma. JAMA Oncol. 2019;5:1141–9.
45. https://clinicaltrials.gov/ct2/show/NCT03740165

46. Pujade-Lauraine E, Fujiwara K, Ledermann JA, Oza AM, Kristeleit R, Ray-Coquard IL, et al. Avelumab alone or in combination with chemotherapy versus chemotherapy alone in platinum-resistant or platinum-refractory ovarian cancer (JAVELIN ovarian 200): an open-label, three-arm, randomised, phase 3 study. Lancet Oncol. 2021;22:1034–46.
47. Monk BJ, Colombo N, Oza AM, Fujiwara K, Birrer MJ, Randall L, et al. Chemotherapy with or without avelumab followed by avelumab maintenance versus chemotherapy alone in patients with previously untreated epithelial ovarian cancer (JAVELIN ovarian 100): an open-label, randomised, phase 3 trial. Lancet Oncol. 2021;22:1275–89.
48. Domchek SM, Postel-Vinay S, Im SA, Park YH, Delord JP, Italiano A, et al. Olaparib and durvalumab in patients with germline BRCA-mutated metastatic breast cancer (MEDIOLA): an open-label, multicentre, phase 1/2, basket study. Lancet Oncol. 2020;21:1155–64.
49. https://clinicaltrials.gov/ct2/show/NCT03737643
50. Moore KN, Bookman M, Sehouli J, Miller A, Anderson C, Scambia G, et al. Atezolizumab, bevacizumab, and chemotherapy for newly diagnosed stage III or IV ovarian cancer: placebo-controlled randomized phase III trial (IMagyn050/GOG 3015/ENGOT-OV39). J Clin Oncol. 2021;39:1842–55.
51. Martin-Sabroso C, Lozza I, Torres-Suarez AI, Fraguas-Sanchez AI. Antibody-antineoplastic conjugates in gynecological malignancies: current status and future perspectives. Pharmaceutics. 2021;13:1705.
52. O'Malley DM, Calo CA. Progress in gynecologic cancers with antibody drug conjugates. Curr Oncol Rep. 2021;23:89.
53. Hafeez U, Parakh S, Gan HK, Scott AM. Antibody-drug conjugates for cancer therapy. Molecules. 2020;25:4764.
54. Erickson BK, Zeybek B, Santin AD, Fader AN. Targeting human epidermal growth factor receptor 2 (HER2) in gynecologic malignancies. Curr Opin Obstet Gynecol. 2020;32:57–64.
55. Khongorzul P, Ling CJ, Khan FU, Ihsan AU, Zhang J. Antibody-drug conjugates: a. Review Mol Cancer Res. 2020;18:3–19.
56. Baah S, Laws M, Rahman KM. Antibody-drug conjugates-a tutorial review. Molecules. 2021;26:2943.
57. Gryshkova V, Goncharuk I, Gurtovyy V, Khozhayenko Y, Nespryadko S, Vorobjova L, et al. The study of phosphate transporter NAPI2B expression in different histological types of epithelial ovarian cancer. Exp Oncol. 2009;31:37–42.
58. Wu B, Yu C, Zhou B, Huang T, Gao L, Liu T, et al. Overexpression of TROP2 promotes proliferation and invasion of ovarian cancer cells. Exp Ther Med. 2017;14:1947–52.
59. Thomas LJ, Vitale L, O'Neill T, Dolnick RY, Wallace PK, Minderman H, et al. Development of a novel antibody-drug conjugate for the potential treatment of ovarian, lung, and renal cell carcinoma expressing TIM-1. Mol Cancer Ther. 2016;15:2946–54.
60. Manzano A, Ocana A. Antibody-drug conjugates: a promising novel therapy for the treatment of ovarian cancer. Cancers (Basel). 2020;12:2223.
61. Moore KN, Oza AM, Colombo N, Oaknin A, Scambia G, Lorusso D, et al. Phase III, randomized trial of mirvetuximab soravtansine versus chemotherapy in patients with platinum-resistant ovarian cancer: primary analysis of FORWARD I. Ann Oncol. 2021;32:757–65.
62. O'Malley DM, Matulonis UA, Birrer MJ, Castro CM, Gilbert L, Vergote I, et al. Phase Ib study of mirvetuximab soravtansine, a folate receptor alpha (FRalpha)-targeting antibody-drug conjugate (ADC), in combination with bevacizumab in patients with platinum-resistant ovarian cancer. Gynecol Oncol. 2020;157:379–85.

# Chapter 2
# Carcinogenesis and Personalization in HPV-Associated Precancer Lesions of the Cervix

**Kei Kawana**

**Abstract** Cervical cancer and other HPV-associated cancers are caused by persistent infection with high-risk HPVs, mainly HPV types 16 and 18. Persistent high expression of the *E6* and *E7* oncogenes of high-risk HPV is essential for cell immortalization and transformation into cancer. Therefore, therapeutic agents targeting HPV E6 and E7 have been developed. It is well-known that cervical intraepithelial neoplasia (CIN), a precancerous lesion of cervical cancer, and cervical cancer can be regulated by host immunity. CIN in particular often spontaneously regresses to normal, which is a result of the host's immune response to HPV viral proteins. Therefore, therapeutic vaccines that induce an immune response against HPV have been developed, but have not yet been commercialized. We have been developing therapeutic vaccines against precancerous lesions by applying the mechanism of mucosal immunity. By investigating the antigen expression and immunosuppressive factors involved in the induction of host immunity, we expect to personalize patients who will respond to HPV oncogene-targeting therapy or anti-HPV immunotherapy.

**Keywords** Cervical cancer · Cervical intraepithelial lesion · Human papillomavirus · Immunotherapy · Mucosal immunology · Carcinogenesis · HPV E6 and E7

## 2.1 Molecular Biology and Epidemiology of Human Papillomavirus (HPV) Infection

Human papillomavirus (HPV) is a small double-stranded DNA virus that infects only humans. HPV has eight genes: early genes *E1, E2, E4, E5, E6*, and *E7*, and late genes *L1* and *L2*. However, there are more than 200 genotypes of HPV, which are

K. Kawana (✉)
Department of Obstetrics and Gynecology, Nihon University School of Medicine, Tokyo, Japan
e-mail: kawana.kei@nihon-u.ac.jp



M. Mandai (ed.), *Personalization in Gynecologic Oncology*, Comprehensive Gynecology and Obstetrics, https://doi.org/10.1007/978-981-19-4711-7_2

classified according to the homology of the viral genes [1]. HPV infects the stratified squamous epithelium of the skin and mucosal epithelium, and its life cycle is completed only in the epithelium. In the parabasal cells, E6 and E7 are expressed and promote cell proliferation. In the middle layer, E1 and E2 promote replication of the viral genome. Near the surface, structural proteins L1 and L2 are simultaneously expressed to form the viral capsids. The L1/L2 capsids package the viral genome into new viral particles that are released into the vagina, along with a detachment of the stratified squamous epithelium. In latent infection, the viral DNA is bound to the genome of the host cell by a small amount of E2, and is transferred to dividing cells with the viral DNA during mitosis [2]. This mechanism allows HPV to continue to exist in the mucosal basal layer.

There have been many epidemiological studies on high-risk HPV associated with carcinogenesis. HPV testing is a method of detecting HPV DNA (viral genes) by diffusion amplification of DNA extracted from cervical exfoliated epithelial cells. It is possible to identify more than 30 high-risk HPV types. HPV does not cause viremia; therefore, antibody induction is weak and titer is low. The positive rate of HPV antibodies is about 50–70%, although it is difficult to know exactly. Based on the antibody positivity rate and HPV DNA detection rate, it is estimated that 50%–80% of all women are exposed to HPV at least once in their lifetime [3]. In a prospective study, Ho et al. collected vaginal washes from 608 female students at a university in the USA over a 3-year period and used PCR to detect HPV DNA in the suspended cells. Of the group who were positive for HPV DNA (43% of 608 students), 31% became negative within 6 months, 39% became negative within 6–12 months, 11% became negative within 12–18 months, and eventually about 90% of the HPV-DNA-positive group became negative within 2 years [4]. In Japan, the HPV-positive rate among pregnant women in their 20s is reported to be 20–30%, and the HPV-positive rate in Japan is equal to or higher than that in developed and emerging countries overseas. The HPV DNA test positivity rate among Japanese women by age is highest in teenagers, ranging from 30% to 40%. After that, the DNA positivity rate decreases with age to 20–30% in women in their 20s, 10–20% in women in their 30s, and 5–10% in women in their 40s [5].

## 2.2 Epidemiology of Cervical Cancer

According to the World Health Organization (WHO) announcement at the International Conference on HPV Vaccines held in Geneva in February 1999, the number of HPV carriers worldwide is estimated to increase by 300 million per year. Of these, about 30 million will develop low-grade squamous intraepithelial lesions (LSILs) annually. Lesions more advanced than precancers can be treated; therefore, it is difficult to determine the frequency of cervical cancer in the natural history of the disease, but cases are currently increasing by about 500,000 per year worldwide [6]. The distribution of HPV genotypes in cervical cancer is as follows. In squamous cell carcinoma, HPV16 accounts for about half of the cases, HPV18 is the second most common, accounting for about 10%, followed by HPV45, 31, and 33. In contrast, in adenocarcinoma and adenosquamous carcinoma, HPV18 accounts for about 40% of cases and is detected to the same extent as HPV16 [3]. HPV18 is detected at a high rate in cervical adenocarcinoma. It has been shown that about 70% of all cervical cancers are caused by HPV16 and 18. However, HPV16 accounts for about 20% of HPV detected in women with no cytological abnormalities, which is a different distribution compared with that in cervical cancer. In Japanese data, the detection frequency of HPV16 and 18 in normal cervical cytology is just over 10%, and HPV52 and 58 are predominant [7]. This indicates that HPV16 and 18 have characteristics that predispose them to invasive cancer. According to the International Agency for Research on Cancer, the odds ratio of developing cervical cancer if any HPV type is detected is 158 times higher than if HPV is not detected. The odds of developing cervical cancer are 434 times higher when HPV16 is detected than when it is undetected, and 248 times higher when HPV18 is detected. This clearly indicates a higher risk of cervical cancer when HPV16 and 18 are detected [3].

## 2.3 Molecular Biological Mechanisms of HPV Carcinogenesis and the Potential for Therapeutics Discovery Targeting these Mechanisms

The mechanism of HPV-induced carcinogenesis has been intensively studied since the 1980s using molecular biological techniques (Fig. 2.1). The molecular biological changes in HPV-infected cells correlate with the pathological development of cervical intraepithelial neoplasia (CIN). In particular, the fact that squamous

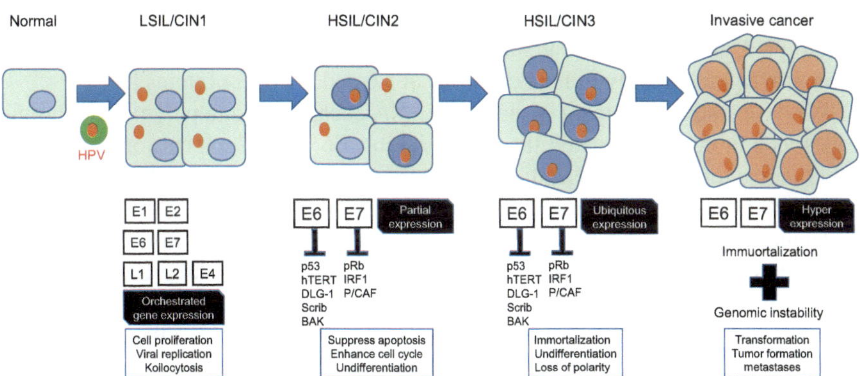


**Fig. 2.1** Carcinogenesis of HPV-associated precancer lesions. Squamous neoplastic lesions of the cervix can now be broadly classified pathologically into low- and high-grade squamous intraepithelial lesions (LSILs and HSILs) may be the result of a better understanding of the steps in HPV carcinogenesis. Interestingly, the molecular biological changes in HPV-infected cells correlate with the pathological development of cervical intraepithelial neoplasia (CIN)

neoplastic lesions of the cervix can now be broadly classified pathologically into low- and high-grade squamous intraepithelial lesions (LSILs and HSILs) may be the result of a better understanding of the steps in HPV carcinogenesis. The LSIL→HSIL→squamous cell carcinoma sequence shown in Fig. 2.1 is a common oncogenic process. In contrast, glandular lesions often progress directly from HPV infection to adenocarcinoma in situ and/or invasive adenocarcinoma. In the Bethesda system of cytological diagnosis, the abnormal finding of atypical gland cells (AGC → adenocarcinoma) can be interpreted as rapid progression.

Here, we discuss the general mechanism of HPV carcinogenesis, and the LSIL shows the morphological change of HPV-infected cells with koilocytosis. In LSILs, HPV expresses E6 and E7 proteins to allow infected cells to proliferate near the parabasal layer (lower third of the epithelium) while maintaining squamous differentiation for viral proliferation. E1 and E2 are expressed for replication of the viral genome and L1 and L2 for the synthesis of the viral particle capsid. These viral proteins are expressed in an orderly fashion as the stratified squamous epithelium differentiates. As mentioned above, HPV can always be retained in the basal cells, and the viral genome is always supplied by the basal cells, so HPV does not disappear by shedding [2]. In HISLs, the orderly expression pattern of the virus is lost in some infected cells, and the cancer proteins E6 and E7 are highly expressed. E6 inactivates such enzymes as p53 and hTERT. E7 inactivates Rb protein and keeps the cell cycle going by activating histone deacetylase and cyclin. These functions inhibit apoptosis, enhance the cell cycle, and inhibit squamous cell differentiation. The result is the transformation of infected cells into immortalized cells. CIN2 can be described as a mixture of virus-infected cells and neoplastic cells, and the neoplastic cells can be distinguished from virus-infected cells by their different histopathological features [8]. In CIN3, immortalized neoplastic cells almost completely replace the epithelial lining, and infected cells and cells with koilocytosis are absent from the epithelium. In this state, the squamous epithelium is no longer differentiated, parabasal-like cells with a high N/C ratio proliferate abnormally, cell polarity is disrupted, and histological architecture is lost. These morphological changes have been proven molecularly to be the actions of E6 and E7 oncoproteins [1].

We have studied the agent of antisense RNA encapsulated in polymeric nanomicelles as a molecular targeted therapy to inhibit E6 and E7 expression by nucleic acid therapeutics [9]. This is a therapy in which siRNA specific for HPV E6 and E7 is administered to inhibit transcription of E6/E7 by RNA interference. The target molecule is an exogenous viral protein; therefore, the effect on normal tissues is low, unlike for ordinary molecular targeted therapies. siRNA-based transcriptional repression of viral genes can be observed in cultured cells in vitro, but there is a problem with drug delivery for human application. Therefore, in collaboration with Kazunori Kataoka and Kanjiro Miyata of the University of Tokyo, using nanotechnology, we have produced a therapeutic agent containing siRNA targeting E6/E7 (E6/E7 siRNA) encapsulated in polymeric micelles [10]. E6/E7 siRNA-encapsulated polymeric micelles were prepared for HPV16 and HPV18. To confirm the inhibitory effect of siRNA on E6/E7 expression in vivo, E6/E7 or control siRNA-encapsulated polymeric micelles were intravenously injected into nude mice transplanted with cervical cancer cells. HPV16 E6/E7 siRNA-encapsulated polymeric micelles suppressed growth of HPV16-positive cervical cancer cell line (SiHa cell) tumors by about 80%, and HPV18 E6/E7 siRNA-encapsulated polymeric micelles suppressed growth of HPV18-positive cervical cancer cell line (HeLa cell) tumors by about 70%, compared with control micelles. The E6/E7 mRNA level in the tumor was significantly decreased in the E6/E7 siRNA polymeric micelle group. Moreover, in the excised tumors, p53 was rescued in a dose-dependent manner by E6/E7 siRNA polymeric micelles [9]. The safety of polymeric nanomicelles in humans has been confirmed for anticancer and granulocyte colony-stimulating factor agents, and the antitumor effect has been confirmed for intravenous administration, which can be used in clinical practice. The manufacturing process for micelles is simple. In the future, there is a possibility that this technology will be developed into a molecular therapy targeting HPV E6/E7.

## 2.4 HPV-Associated Carcinogenesis from the Viewpoint of Morphology and the Possibility of Drug Discovery Targeting These Mechanisms

High-risk HPV-associated cancers have been reported to include cervical, anal, vulvar, vaginal, and oropharyngeal cancers, of which, cervical cancer is caused by HPV with most highly rate (about 95%). About 70% of cervical cancer and about 90% of anal, vulvar, vaginal, and oropharyngeal cancers are caused by HPV16 or 18. Most HPV-associated carcinogenesis outside the cervix is caused by HPV16 and 18, and HPV16 and 18 are by far the most likely HPV types to cause cancer [3]. In other words, the cervix is the only organ that is uniquely susceptible to cancer caused by high-risk HPVs other than HPV16 or 18. High-risk HPV, in contrast, infects every part of the external genitalia: cervix, vagina, and vulva. Why is cancer more likely to develop in the cervix than in the other external genitalia? These questions can be answered as follows. One of the reasons is the presence of the squamocolumnar junction (SCJ) at the cervix. Tissue stem cells called reserve cells are localized in the SCJ. Reserve cells are referred to as stem cells because they have the pluripotency to differentiate into squamous and glandular epithelium, and have the ability to self-renew. HPV infection and CIN tend to occur in the SCJ or transformation zone, which means that HPV has a propensity to infect self-renewing, pluripotent cells in the SCJ. The monolayer structure of SCJ cells allows for easy entry without the need for deep wounds. SCJ cells also have the ability to differentiate by squamous stratification (squamous metaplasia), which is thought to be induced by HPV [11] (Fig. 2.2a). HPV is able to replicate itself through stratified squamous epithelium, and its dormancy and ability to self-renew allow for repeated latency and viral proliferation. The SCJ is an ideal place for HPV to infect.

When tissue stem cells in the SCJ undergo malignant transformation through HPV carcinogenesis, they have the properties of cancer stem cells (Fig. 2.2b). As a result, they are likely to develop quickly and highly into invasive cancer cells. One of

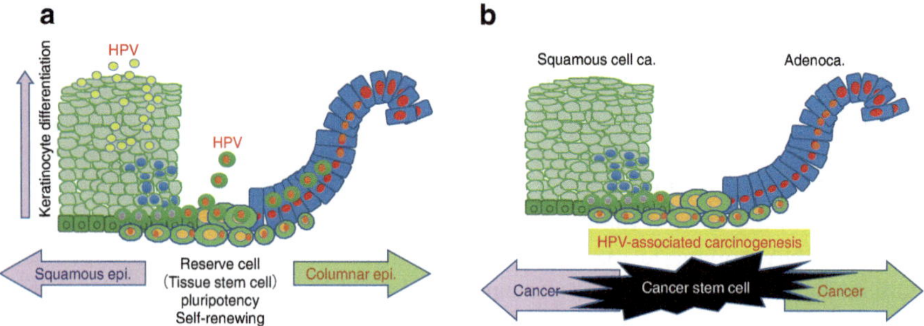


**Fig. 2.2** HPV-associated tissue stem cell (reserve cell) and cancer stem cell in the cervix. (**a**) Tissue stem cells called reserve cells are localized in the squamocolumnar junction (SCJ) at the cervix and referred to as stem cells because they have the pluripotency to differentiate into squamous and glandular epithelium, and have the ability to self-renew. SCJ cells also have the ability to differentiate by squamous stratification (squamous metaplasia), which is thought to be induced by HPV. (**b**) When tissue stem cells in the SCJ undergo malignant transformation through HPV carcinogenesis, they have the properties of cancer stem cells. HPV-infected reserve cells are pluripotent, meaning development of squamous cell carcinoma and adenocarcinoma. Stemness of the cancer cells have been identified as a specific mechanism of HPV carcinogenesis in cervical cancer

the reasons why cervical cancer occurs at a young age may be that cervical cancer arises from the SCJ, which has cells that have stem cell features. Furthermore, HPV-infected reserve cells are pluripotent, meaning that a change in the squamous cells results in the development of squamous cell carcinoma, and a change in the glandular cells leads to development of adenocarcinoma. Stem cells and cancer stem cells have been identified as a specific mechanism of HPV carcinogenesis in cervical cancer; therefore, development of therapeutics targeting these stem cell features is expected.

We have succeeded in artificially generating reserve cells as stem cells in the SCJ, and have established a technique to differentiate induced pluripotent stem cells through intermediate mesoderm cells to reserve-cell-like cells called induced reserve cells (iRCs). iRCs are pluripotent cells that can differentiate into both glandular and squamous epithelium in 3D culture and are positive for stem cell markers, Mullerian duct markers, and SCJ markers, making them reserve cell-like [12]. We also have generated iRCs with E6 and E7 oncogenes of HPV16 and 18, and are performing RNA sequencing analysis. By analysis of HPV16 and 18 E6/E7-positive iRCs that maintain stemness, we are exploring new targets based on the gene expression profiles that show characteristics of cancer stem cells.

## 2.5 Drug Discovery and Development of Immunotherapy Targeting HPV-Associated Cancers

E6 and E7 proteins of HPV are both initiators and promoters of carcinogenesis from HPV infection to cervical cancer (Fig. 2.1). E6 and E7 are ubiquitously expressed in cervical cancer, precancer, and other HPV-associated cancers, and are necessary for maintenance of malignant transformation. E6 and E7 are immunogenic in humans and are presented as antigens on the surface of cervical cancer and precancer cells, making them so-called cancer antigens. Furthermore, because they are viral proteins, they do not affect normal host cells. HPV E6 and E7 oncoproteins are cancer antigens specific to HPV-associated cancers and are promising target molecules for immunotherapy. Immunotherapies targeting these proteins have been developed over the past two decades; however, none is clinically applicable.

Recently, the expression of immune checkpoints, programmed death (PD)-1/PD ligand (PD-L)1, and MSI-high tumors in cervical cancer have been found, and immunotherapy using immune checkpoint inhibitors (ICIs) has been reconsidered [13]. PD-L1 is upregulated by HPV E5, E6, and E7, and expression of PD-1 and PD-L1 in cervical cancer or CIN is enhanced [14–16]. ICIs, anti-PD-1 and anti-PD-L1, will be combined with HPV E6/E7-targeted cancer immunotherapy and these biomarkers may be used for personalization in HPV-associated cancer or precancer lesions.

## 2.6 Reason for Promising Immunotherapy for Precancer Lesions CIN2/3

The only treatment available for early cervical cancer and its precancerous lesions, which peak in women in their 20s and 30s, is surgery. At present, there is no treatment available for CIN2/3. Total hysterectomy terminates fertility, and conization is associated with poor perinatal outcomes of subsequent pregnancies. The preterm birth rate increases about threefold in pregnancies after conization, as do the rates of cesarean delivery and low birth weight [17]. The age of patients with CIN3 often coincides with the age at which they become pregnant and give birth; therefore, the increased rate of perinatal complications caused by treatment is a major problem.

In HPV-associated precursor lesions, including precancer, spontaneous regression induced by the host immune response to HPV proteins is often observed because of viral carcinogenesis. Many cohort studies of CIN lesions have demonstrated that within 2 years of follow-up, about 60% of CIN2 and 20% of CIN3 spontaneously regress [18, 19]. This is thought to be a result of the spontaneous induction of host cellular immune responses, mainly against E6 and E7 proteins, in patients with CIN2/3. Research on the application of this natural regression to immunotherapy with therapeutic vaccines has been initiated worldwide since the 1990s.

A large number of clinical trials (Phase I–III) have been conducted on HPV-targeted immunotherapy in CIN2/3. Molecular targets of these studies are mostly the E7 or E6/E7 proteins [20]. In prior clinical trials, vaccine antigen was administered by intramuscular or subcutaneous injection to induce E6/E7-specific cell-mediated immunity in the peripheral blood. However, the immune response does not always correlate with clinical efficacy, and none of the therapeutic vaccines has been commercialized at this time. The most advanced therapeutic vaccine currently in development is VGX-3100 (Inovio Inc.), which is a plasmid DNA vaccine that is injected intramuscularly [21]. In a phase IIb study of 167 patients with HPV16-positive CIN2/3, the rate of histopathological regression and viral clearance was 40% in the VGX-3100 group and 12% in the placebo group in modified intention to treat analysis. Although there was a significant difference in efficacy, inoculation site adverse events occurred in 98% of patients. In addition, the website reports the results of a phase III study of 201 patients with CIN2/3 as a modified intention to treat population. The primary endpoint was a histologically confirmed LSIL/normal that was negative for HPV16 and/or HPV18 DNA at week 36. The percentage of patients with LSIL/normal and viral clearance was 22.5% with VGX-3100 and 11.1% with placebo, which was not significantly different. Comparison of the primary endpoint results of the phase IIb and III studies showed that the efficacy of VGX-3100 was lower in the phase III trial, although the results in the placebo group were similar. At this point, no promising therapeutic candidates for CIN2/3 have been developed.

## 2.7 Development of Mucosal Immunotherapy for CIN2/3 Based on Histopathogenesis of CIN2/3

We focused on the fact that CIN2/3 is a mucosal lesion, and developed immunotherapy using the mucosal immune system (called mucosal immunotherapy). In mucosal immunity of the cervical epithelium, gut-associated lymphoid tissue (GALT), including Peyer's patches or mesenteric lymph nodes in the intestinal mucosa, is an organized inductive tissue (Fig. 2.3). Mucosal T cell precursors are primed to produce antigen-specific helper and killer T cells, and are imprinted for

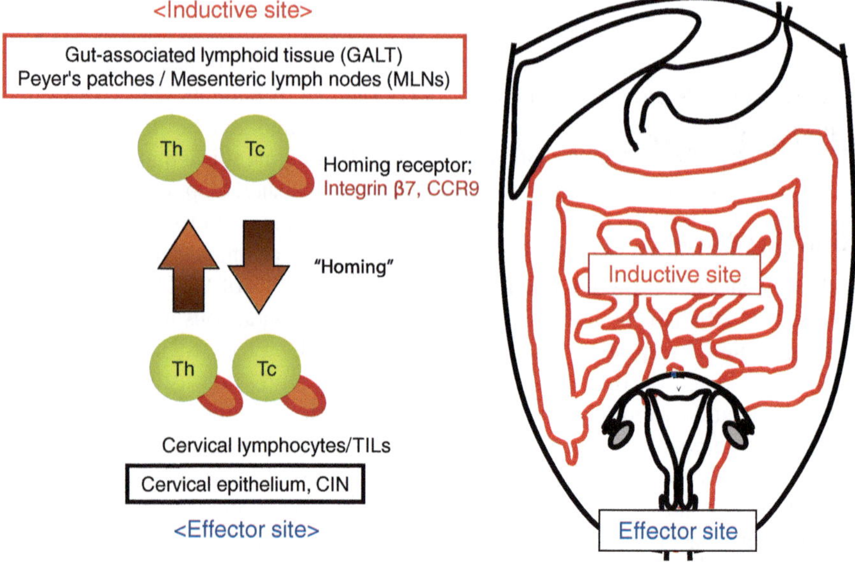


**Fig. 2.3** Mucosal immune system in the cervix. In mucosal immunity of the cervix, gut-associated lymphoid tissue (GALT), including Peyer's patches or mesenteric lymph nodes in the intestinal mucosa, is an organized inductive tissue. Mucosal T cell precursors are primed to produce antigen-specific helper and killer T cells, and are imprinted for homing receptors (integrin β7 and CCR9) and T helper (Th)1/Th2 polarization by dendritic cells in GALT. The primed and memory T cells "homing" to the effector sites are activated by antigen stimulation in the mucosal lesions of the cervix

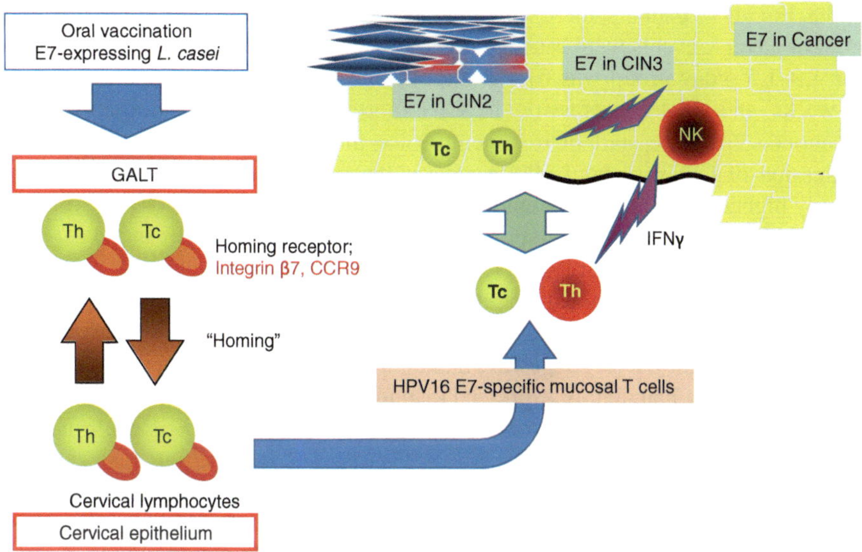


**Fig. 2.4** Putative pharmacological effects of oral vaccination with E7-expressing *L. casei*. The orally administered E7-expressing *Lactobacillus*-based therapeutic vaccine are taken up by GALT from the intestinal tract. Mucosal T cells are primed to produce E7-specific immune cells in GALT and home to the cervical mucosa; E7 expressed in CIN2–3 is recognized and TH1 immune responses are elicited, resulting in antitumor effects including NK activity through TH1 cytokines

homing receptors (integrin β7 and CCR9) and T helper (Th)1/Th2 polarization by dendritic cells in GALT. The primed and memory T cells homing to the effector sites are activated by antigen stimulation in the mucosal lesions of the cervix.

Gut-derived mucosal T cells infiltrate CIN as tumor-infiltrating lymphocytes. We found that 20s–40% of the lymphocytes present in the cervical epithelium of patients with CIN were gut-derived integrin $\beta7^+$ and $CCR9^+$ T cells, and CIN was more likely to regress to normal when their content was high [22]. Thus, infiltration of integrin $\beta7^+$ and $CCR9^+$ T cells (especially Th1 T cells) primed and imprinted in the intestinal mucosa contributes to spontaneous regression of CIN.

We hypothesize that by administering cancer antigen HPV E7 to the intestinal mucosa, inducing Th1-type mucosal immunity against E7 by GALT, and homing gut-derived mucosal T cells to the cervical epithelium, E7-specific mucosal Th1 cells recognizing E7-expressing CIN2/3 produce interferon-γ and Th1 responses, including NK cell activity, leading to antitumor activity (Fig. 2.4). We generated *Lactobacillus casei* expressing HPV16 E7 (code name: GLBL101c) as a vaccine antigen for cancer immunotherapy. We confirmed the killer activity of mucosal T cells against HPV16 E7-expressing epithelial (TC-1) cells in preclinical studies in mice [23]. After this, a proof-of-concept clinical study was conducted at the University of Tokyo Hospital under the approval of the Institutional Review Board. The patients had histopathologically confirmed CIN3 that was positive for HPV16 alone, and received GLBL101c orally once a day for 5 days/week for weeks 1, 2, 4, and 8. For all 17 patients, there were no adverse events of grade 2 or higher, and no grade 1 adverse events were causally related. The clinical efficacy of GLBL101c in the 1.0–1.5 g/day group was 61.5% regression to CIN2 (partial response; PR) at 9 weeks and 38.4% regression to CIN1/normal (complete response; CR) at 12 months from the start of treatment. Since the estimated rate of spontaneous regression of CIN3 is about 10% per year, the regression rate of our study was clearly higher than that of spontaneous regression. In addition, the group with CR/PR had higher induction of E7-specific interferon-γ-producing cells in the cervical epithelium than the nonregressed group had [24].

We conducted a randomized, double-blind, placebo-controlled trial of GLBL101c in CIN2 patients who were positive for HPV16 alone. However, the results showed limited efficacy for CIN2, suggesting higher efficacy may be expected for CIN3, in which E7 is more abundantly expressed than in CIN2 [25]. Compared with the placebo group, there was no difference in adverse events, and safety was confirmed by the second clinical study.

The first-generation, E7-expressing, *Lactobacillus*-based therapeutic vaccine, GLBL101c, used in these two exploratory clinical studies was considered to have limited pharmacological efficacy. Therefore, we developed a next-generation agent (code name: IGMKK16E7) with several times higher E7-specific Th1 immune responses [26]. Then, a phase I/II investigator-initiated clinical trial of IGMKK16E7 was conducted. This was an intergroup, parallel, randomized placebo-controlled trial (mucosal immunotherapy using *Lactobacillus* for treatment of squamous intraepithelial lesion: MILACLE trial) of four groups: placebo, low-dose, medium-dose, and high-dose IGMKK16E7 in HPV16-positive CIN2/3. The target number of enrolled patients was 164 (124 with CIN3 and 40 with CIN2). The primary endpoint was histopathological regression to normal (CR) or CIN1 (PR) at week 16 after the start of treatment [27]. This clinical trial has already finished enrolling patients, and the final analysis is scheduled to be published in Summer 2022.

For immunotherapy of HPV-associated cancer or precancer lesions, expression of viral oncoproteins and/or immunosuppressive biomarkers, including immune checkpoints, could be used to personalize treatment. It is expected that the immune response (pharmacological effect) will differ among patients depending on the expression level of the target molecule of immunotherapy. CIN2/3 lesions with higher E7 expression levels are expected to induce a stronger E7-specific Th1 response, resulting in a higher therapeutic effect. If the expression of immune checkpoints such as PD-L1 is increased by HPV E6/E7, immunosuppression may be enhanced. In this case, a combination of ICIs may be effective. Thus, the indications for immunotherapy and combination with ICIs can be personalized according to the characteristics of each CIN2/3 patient.

**Acknowledgments** We are grateful to Mrs. Naoko Tomita for her support; and to GLOVACC Inc. for GLBL101c, IGMKK16E7, and placebo. This research was supported by the Agency for Medical Research and Development (AMED), Japan. We thank Mr. Cathel Kerr from Edanz (https://jp.edanz.com/ac) for editing a draft of this manuscript.

**Conflicts of Interest** GLOVACC Inc. gifted GLBL101c, IGMKK16E7, and placebo, and partially supported our clinical trial.

## References

1. Zur Hausen H. Papillomavirus and cancer: from basic studies to clinical application. Nat Rev Cancer. 2002;2:342–50.
2. You J. Papillomavirus interaction with cellular chromatin. Biochim Biophys Acta. 2010;1799(3–4):192–9.
3. Bosch FX, et al. Human papillomavirus and cervical cancer – burden and assessment of causality. J Natl Cancer Inst Monogr. 2003;31:3–13.
4. Ho GYF, et al. Natural history of cervicovaginal papillomavirus infection in young women. N Engl J Med. 1998;338:423.
5. Inoue M, Sakaguchi J, Sasagawa T, Tango M. The evaluation of human papillomavirus DNA testing in primary screening for cervical lesions in a large Japanese population. Int J Gynecol Cancer. 2006;16:1007–13.
6. The current status of development of prophylactic vaccines against human papillomavirus infection. Report of a technical meeting. Geneva, 16–18 February, 1999.
7. Miura S, et al. Do we need a different strategy for HPV screening and vaccination in East Asia? Int J Cancer. 2006;119:2713–5.
8. Furuta R, Hirai Y, Katase K, Tate S, Kawaguchi T, Akiyama F, Kato Y, Kumada K, Iwasaka T, Yaegashi N, Kanazawa K, Yoshikawa H, Kitagawa T. Ectopic chromosome around centrosome in metaphase cells as a marker of high-risk human papillomavirus-associated cervical intraepithelial neoplasias. Int J Cancer. 2003;106:167–71.
9. Nishida H, Matsumoto Y, Kawana K, Christie RJ, Naito M, Kim BS, Toh K, Min HS, Yi Y, Matsumoto Y, Kim HJ, Miyata K, Taguchi A, Tomio K, Yamashita A, Inoue T, Nakamura H, Fujimoto A, Sato M, Yoshida M, Adachi K, Arimoto T, Wada-Hiraike O, Oda K, Nagamatsu T, Nishiyama N, Kataoka K, Osuga Y, Fujii T. Systemic delivery of siRNA by actively targeted polyion complex micelles for silencing the E6 and E7 human papillomavirus oncogenes. J Control Release. 2016;231:29–37.
10. Christie RJ, Matsumoto Y, Miyata K, Nomoto T, Fukushima S, Osada K, Halnaut J, Pittella F, Kim HJ, Nishiyama N, Kataoka K. Targeted polymeric micelles for siRNA treatment of experimental cancer by intravenous injection. ACS Nano. 2012;6(6):5174–89.
11. Mirkovic J, Howitt BE, Roncarati P, Demoulin S, Suarez-Carmona M, Hubert P, McKeon FD, Xian W, Li A, Delvenne P, Crum CP, Herfs M. Carcinogenic HPV infection in the cervical squamo-columnar junction. J Pathol. 2015;236:265–71.
12. Sato M, Kawana K, Adachi K, Fujimoto A, Yoshida M, Nakamura H, Nishida H, Inoue T, Taguchi A, Ogishima J, Eguchi S, Yamashita A, Tomio K, Wada-Hiraike O, Oda K, Nagamatsu T, Osuga Y, Fujii T. Regeneration of cervical reserve cell-like cells from human induced pluripotent stem cells (iPSCs): a new approach to finding targets for cervical cancer stem cell treatment. Oncotarget. 2017;8:40935–45.
13. Garbuglia AR, Lapa D, Sias C, Capobianchi MR, Del Porto P. The use of both therapeutic and prophylactic vaccines in the therapy of papillomavirus disease. Front Immunol. 2020 Feb;18(11):188. https://doi.org/10.3389/fimmu.2020.00188.
14. Liu C, Lu J, Tian H, Du W, Zhao L, Feng J, et al. Increased expression of PDL1 by the human papillomavirus 16 E7 oncoprotein inhibits anticancer immunity. Mol Med Rep. 2017;15:1063–70. https://doi.org/10.3892/mmr.2017.6102.
15. Mezache L, Paniccia B, Nyinawabera A, Nuovo GJ. Enhanced expression of PD L1 in cervical intraepithelial neoplasia and cervical cancers. Mod Pathol. 2015;28:1594–602. https://doi.org/10.1038/modpathol.2015.108.
16. Yang W, Song Y, Lu YL, Sun JZ, Wang HW. Increased expression of programmed death (PD)-1 and its ligand PD-L1 correlates with impaired cell-mediated immunity in high-risk human papillomavirus related cervical intraepithelial neoplasia. Immunology. 2013;139:513–22. https://doi.org/10.1111/imm.12101.
17. Kyrgiou M, Koliopoulos G, Martin-Hirsch P, et al. Obstetric outcomes after conservative treatment for intraepithelial or early invasive cervical lesions: systematic review and meta-analysis. Lancet. 2006;367:489–98.

18. Matsumoto K, Oki A, Furuta R, Maeda H, Yasugi T, Takatsuka N, Mitsuhashi A, Fujii T, Hirai Y, Iwasaka T, Yaegashi N, Watanabe Y, Nagai Y, Kitagawa T, Yoshikawa H. Japan HPV and cervical cancer study group. Int J Cancer. 2011;128:2898–910.
19. Holowaty P, Miller AB, Rohan T, To T. Natural history of dysplasia of the uterine cervix. J Natl Cancer Inst. 1999;91:252–8.
20. Kawana K, Yasugi T, Taketani Y. Human papillomavirus vaccines: current issues and future: review. Indian J Med Res. 2009;130:341–7.
21. Trimble CL, Morrow MP, Kraynyak KA, Shen X, Dallas M, Yan J, Edwards L, Parker RL, Denny L, Giffear M, Brown AS, Marcozzi-Pierce K, Shah D, Slager AM, Sylvester AJ, Khan A, Broderick KE, Juba RJ, Herring TA, Boyer J, Lee J, Sardesai NY, Weiner DB, Bagarazzi ML. Safety, efficacy, and immunogenicity of VGX-3100, a therapeutic synthetic DNA vaccine targeting human papillomavirus 16 and 18 E6 and E7 proteins for cervical intraepithelial neoplasia 2/3: a randomised, double-blind, placebo-controlled phase 2b trial. Lancet. 2015;386:2078–88.
22. Kojima S, Kawana K, Fujii T, Yokoyama T, Miura S, Tomio K, Tomio A, Yamashita A, Adachi K, Sato H, Nagamatsu T, Schust DJ, Kozuma S, Taketani Y. Characterization of gut-derived intraepithelial lymphocyte (IEL) residing in human papillomavirus (HPV)-infected intraepithelial neoplastic lesions. Am J Reprod Immunol. 2011;66:435–43.
23. Adachi K, Kawana K, Yokoyama T, et al. Oral immunization with lactobacillus casei vaccine expressing human papillomavirus (HPV) type 16 E7 is an effective strategy to induce mucosal cytotoxic lymphocyte against HPV16 E7. Vaccine. 2010;28:2810–7.
24. Kawana K, Adachi K, Kojima S, Taguchi A, Tomio K, Yamashita A, Nishida H, Nagasaka K, Arimoto T, Yokoyama T, Wada-Hiraike O, Oda K, Sewaki T, Osuga Y, Fujii T. Oral vaccination against HPV E7 for treatment of cervical intraepithelial neoplasia grade 3 (CIN3) elicits E7-specific mucosal immunity in the cervix of CIN3 patients. Vaccine. 2014;32:6233–9.
25. Ikeda Y, Adachi K, Tomio K, Eguchi-Kojima S, Tsuruga T, Uchino-Mori M, Taguchi A, Komatsu A, Nagamatsu T, Oda K, Kawana-Tachikawa A, Uemura Y, Igimi S, Osuga Y, Fujii T, Kawana K. A placebo-controlled, double-blind randomized (phase IIB) trial of oral administration with HPV16 E7-expressing *lactobacillus*, GLBL101c, for the treatment of cervical intraepithelial neoplasia grade 2 (CIN2). Vaccines (Basel). 2021;9:329.
26. Komatsu A, Igimi S, Kawana K. Optimization of human papillomavirus (HPV) type 16 E7-expressing lactobacillus-based vaccine for induction of mucosal E7-specific IFNγ-producing cells. Vaccine. 2018;36(24):3423–6.
27. Ikeda Y, Uemura Y, Asai-Sato M, Nakao T, Nakajima T, Iwata T, Akiyama A, Satoh T, Yahata H, Kato K, Maeda D, Aoki D, Kawana K. Safety and efficacy of mucosal immunotherapy using human papillomavirus (HPV) type 16 E7-expressing *lactobacillus*-based vaccine for the treatment of high-grade squamous intraepithelial lesion (HSIL): the study protocol of a randomized placebo-controlled clinical trial (MILACLE study). Jap J Clin Oncol. 2019;49:877–80.

# Chapter 3
# Personalized Treatment for Gestational Trophoblastic Neoplasia

**Kazuhiko Ino**

**Abstract** Gestational trophoblastic neoplasia (GTN) arises from abnormal/neoplastic placental trophoblasts, and comprises a spectrum of premalignant invasive hydatidiform moles to malignant tumors, including gestational choriocarcinoma, and two rare types of GTN such as placental site trophoblastic tumor (PSTT) and epithelioid trophoblastic tumor (ETT). As GTN is generally highly chemo-sensitive and patients with GTN frequently desire the preservation of fertility, chemotherapy is performed as an initial treatment in the majority of cases. Based on the FIGO risk scoring system, GTN is classified as low-risk GTN and high-risk GTN, and single-agent chemotherapy is recommended for the former and multi-agent chemotherapy is recommended for the latter. On the other hand, surgery is recommended as the initial treatment for PSTT and ETT. Although standard therapy results in a high survival rate, some (approximately 10%) patients with advanced/metastatic high-risk GTN or PSTT/ETT exhibit chemo-resistance, and the prognosis of these patients is poor. Recently, much attention has been paid to the use of immune checkpoint inhibitors, such as anti-PD-1 or anti-PD-L1 antibodies, as a single therapy or in combination with antiangiogenic therapy, with high clinical efficacy. Although further accumulation of evidence from clinical trials and investigation of biomarkers for good responders to immunotherapy is needed, individualized treatment using immune checkpoint inhibitors may be effective for chemo-resistant/refractory GTN.

**Keywords** Gestational trophoblastic neoplasia (GTN) · Choriocarcinoma · Invasive hydatidiform mole · Placental site trophoblastic tumor (PSTT) · Epithelioid trophoblastic tumor (ETT) · Chemotherapy · Immunotherapy · Immune checkpoint inhibitor

K. Ino (✉)
Department of Obstetrics and Gynecology, Wakayama Medical University, Wakayama, Japan
e-mail: kazuino@wakayama-med.ac.jp



M. Mandai (ed.), *Personalization in Gynecologic Oncology*, Comprehensive Gynecology and Obstetrics, https://doi.org/10.1007/978-981-19-4711-7_3

## 3.1 Introduction

Gestational trophoblastic disease is a general term for diseases that arise from trophoblasts that constitute the placenta during pregnancy and exhibit abnormal proliferation/neoplastic changes. It is classified into five entities: hydatidiform mole, invasive mole, gestational choriocarcinoma, placental trophoblastic tumor (PSTT), and epithelioid trophoblastic tumor (ETT) [1, 2]. Among these diseases, hydatidiform mole is considered abnormal pregnancy caused by abnormality of the fertilization mechanism rather than a neoplasm. The other four diseases are termed gestational trophoblastic neoplasia (GTN) according to the international classification by the International Federation of Gynecology and Obstetrics (FIGO) because they have neoplastic features, and are clinically managed and treated as neoplastic lesions. However, the latest WHO2020 classification [3] classifies invasive moles as a type of molar pregnancy from a pathological viewpoint, whereas gestational choriocarcinoma, PSTT, ETT, and mixed trophoblastic tumor are malignant neoplasms.

Due to advances in chemotherapy and techniques for the measurement of human chorionic gonadotropin (hCG) in blood since the 1980s, the prognosis of GTN has markedly improved, and survival rates of nearly 100% and approximately 90% have been achieved for low- and high-risk GTN, respectively. Furthermore, because of the recent decreases in the number of pregnancies and deliveries and the establishment of management after molar pregnancies, the incidence of GTN has decreased, and the disease is now regarded as a rare tumor. However, as some patients decline treatment and occasionally die, new personalized treatment is considered necessary in addition to multidisciplinary therapy based primarily on conventional chemotherapy for further improvement of the prognosis. Recently, a series of reports on the high efficacy of immune checkpoint inhibitors against treatment-resistant GTN has gained attention. In this review, (1) risk classification of GTN according to a scoring system and treatment based on it, (2) histopathological and immunohistochemical differential diagnosis of PSTT and ETT and therapeutic strategy, and (3) evidence of the efficacy of immune checkpoint inhibitors against GTN and the potential of future immunotherapy are discussed.

## 3.2 Diagnosis and Treatment for Low- and High-Risk GTN

### *3.2.1 Diagnosis of GTN by Scoring System*

As GTN originates from placental trophoblasts, it develops after a preceding pregnancy. Invasive moles occur following 15–20% of complete hydatidiform moles and 0.5–2% of partial hydatidiform moles, the interval from the previous pregnancy to the onset is often within 6 months, and the disease is usually diagnosed during follow-up of serum hCG after removal of the hydatidiform mole. On the other hand, gestational choriocarcinoma may originate from all pregnancies, including term

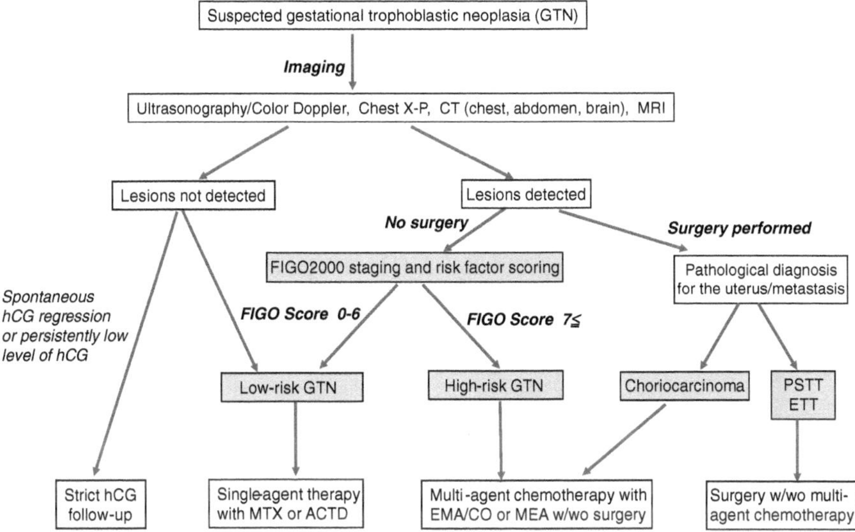


**Fig. 3.1** Flowchart of diagnosis and treatment for gestational trophoblastic neoplasia. On the basis of FIGO2000 risk factor scoring, single-agent therapy is selected for low-risk GTN, whereas multi-agent chemotherapy is selected for high-risk GTN and pathologically diagnosed choriocarcinoma. Surgery and multi-agent chemotherapy are selected for pathologically diagnosed PSTT or ETT

delivery and abortion, and hydatidiform moles. As tumor cells derived from trophoblasts secrete hCG in both conditions, serum hCG is a useful diagnostic and therapeutic marker.

The flowchart of the diagnosis and treatment for GTN is shown in Fig. 3.1. If GTN is suspected based on the clinical course or symptoms or changes in the serum hCG level after a previous pregnancy, imaging examinations are performed first. Ultrasound tomography (combined with color Doppler) and magnetic resonance imaging (MRI) of the pelvis are useful for detecting uterine lesions. Contrast-enhanced computed tomography (CT) is used to detect metastatic lesions. Lung metastases are observed in approximately one-third of patients with invasive moles or vaginal metastasis in rare cases. However, lung metastasis is observed in approximately two-thirds of patients with choriocarcinoma, possibly with metastases to the liver and brain. Therefore, CT of the chest, abdomen, and head is necessary. As GTN occurs in the 20s–40s, many patients wish to retain fertility. Moreover, as the disease responds to chemotherapy, most patients select chemotherapy as the first-line treatment, in which case, a pathological diagnosis is not obtained. For this reason, if a lesion is detected by imaging examinations, the type of antecedent pregnancy, interval between pregnancy and onset, serum hCG level, lesion size, and site and number of metastatic lesions are rated according to the FIGO2000 staging and risk factor scoring system [4], and lesions with a score of 0–6 points and those with a score of 7 points or above are respectively classified as low- and high-risk

GTN. The former mostly corresponds to invasive moles and the latter to gestational choriocarcinoma. If the primary lesion of the uterus or metastatic lesion is treated surgically, the disease may be pathologically diagnosed. On the other hand, pathological examination is essential for the diagnosis of PSTT and ETT, which are rare GTN originating from intermediate-type trophoblasts, and risk factor scoring is not applied. In addition, if the serum hCG level demonstrates a slight natural decrease or remains low while lesions cannot be detected by imaging examinations, careful follow-up may be selected. Based on the clinical or pathological diagnosis according to the scoring method described above, GTN is classified as in the lower part of the flowchart, and the strategy for first-line treatment is decided (Fig. 3.1).

### 3.2.2 Standard Treatments for GTN

Standard treatments for GTN by dividing cases into three groups are summarized in Table 3.1. Refer to NCCN guidelines [5] or ESMO guideliens [6] for details of the dosage and regimen of individual chemotherapies. Single-agent therapy using methotrexate (MTX) or actinomycin D (ACTD) is recommended as the first-line treatment for low-risk GTN, including invasive moles [5, 6]. The remission rate after first-line treatment is 60–80%. In patients resistant to first-line treatment or who exhibit serious adverse events to first-line MTX, MTX is changed to ACTD, etoposide (ETP) alone, or a multi-agent combination (e.g., EMA/CO) is adopted as the second-line treatment. After the serum hCG level decreases below the cutoff level (0.2–2.0 mIU/mL), 1–3 courses of additional chemotherapy are necessary.

**Table 3.1** Standard treatment for gestational trophoblastic neoplasia (GTN)

| *Low-risk GTN* | |
|---|---|
| First line | • Single-agent therapy with methotrexate (MTX) or actinomycin D (ACTD) |
| Second line | • Change from MTX to ACTD (or from ACTD to MTX)<br>• Single-agent therapy with etoposide (ETP)<br>• Multi-agent chemotherapy (EMA/CO) |
| *High-risk GTN*[a] | |
| First line | • Multiagent chemotherapy with EMA/CO or MEA |
| Second line | • EP/EMA (ETP/Cisplatin(CDDP) + ETP/MTX/ACTD)<br>• TP/TE (paclitaxel/CDDP + paclitaxel/ETP)<br>• Surgery for chemoresistant lesions, stereotactic radiotherapy for brain lesions |
| PSTT/ETT | • Hysterectomy, surgical resection for metastatic lesions |
| | • Multiagent chemotherapy with EP/EMA |

*MTX* Methotrexate, *ACTD* Actinomycin D, *ETP* Etoposide, *EMA/CO* Etoposide + Methotrexate + Actinomycin D / Cyclophosphamide + Vincristine, *MEA* Methotrexate + Etoposide + Actinomycin D, *EP/EMA* Etoposide + Cisplatin / Etoposide + Methotrexate + Actinomycin D, *TP/TE* Paclitaxel + Cisplatin / Paclitaxel + Etoposide

[a]Including pathologically-diagnosed choriocarcinoma

Patients who wish to have children are allowed to have the next pregnancy if the serum hCG level remains normal for 1 year after remission of GTN. The survival rate of low-risk GTN patients is nearly 100%.

Next, as the first-line treatment for high-risk GTN, including pathologically diagnosed choriocarcinoma, multi-agent chemotherapy, such as EMA/CO (ETP, MTX, ACTD, cyclophosphamide, vincristine) or MEA (MTX, ETP, ACTD), is recommended [5, 6]. The remission rate by the first-line treatment is approximately 80%. As the second-line treatment, EP/EMA including cisplatin or TP/TE including paclitaxel is used [5, 6]. After the serum hCG level decreases below the cutoff level, 3–4 courses of additional chemotherapy are performed, and a judgment of remission is made if no increase in hCG is observed. The survival rate of high-risk GTN patients is approximately 90%, and the disease becomes resistant to chemotherapy and refractory in approximately 10% of patients, with some cases resulting in death.

Indications of surgery for high-risk GTN are limited, but surgical removal is selected if bleeding is difficult to control, or for uterine lesions or isolated metastatic lesions resistant to chemotherapy [5, 6]. Surgery may be performed for brain metastases if there is a disturbance of consciousness or symptoms of intracranial hypertension. Stereotactic radiotherapy, such as the γ-knife, may also be selected for brain metastases [5, 6].

## 3.3 Diagnosis and Treatment for PSTT and ETT

### *3.3.1 Origins and Histological Characteristics of PSTT and ETT*

PSTT and ETT are both rare GTN that originate from intermediate-type trophoblasts, and histopathological evidence is indispensable for their diagnosis [7–9]. Both tumors develop after pregnancy, including hydatidiform moles and term delivery. Tumor cells of PSTT and ETT have a lower ability to secrete hCG than those of choriocarcinoma, and the serum hCG level is generally low. Although choriocarcinoma responds to chemotherapy, PSTT and ETT are usually poorly responsive to chemotherapy, thus the pathological differentiation of the three diseases is important. Histologically, tumor cells of choriocarcinoma, which resemble two types of trophoblasts, i.e., syncytiotrophoblasts and cytotrophoblasts, form a 2-cell pattern, and, with the addition of tumor cells that resemble intermediate-type trophoblasts, they exhibit a characteristic 3-cell pattern lacking chorionic morphology. In contrast, PSTT characteristically demonstrates monotonous proliferation and myometrial invasion of round or fusiform tumor cells resembling implantation site intermediate trophoblasts, with rich and mildly eosinophilic or clear cytoplasm and infiltration of tumor cells thrusting into gaps among uterine smooth muscle fiber bundles [7]. On the other hand, in ETT, tumor cells resembling chorionic-type intermediate trophoblasts derived from the chorion laeve in the egg membrane proliferate characteristically, exhibiting nest-like, cord-like, and geographical patterns, and invade while maintaining the epithelioid morphology [8].

### 3.3.2 Differential Diagnosis of GTN by Immunohistochemical Staining

The points of differentiation of choriocarcinoma, PSTT, and ETT by immunostaining are summarized in Table 3.2. All three tumors are positive for cytokeratin (CK), which is an epithelial cell marker, positive for inhibin-α, which indicates the trophoblast origin, and generally positive for placental alkaline phosphatase. Choriocarcinoma is strongly positive, but PSTT and ETT are focally positive or negative, for hCG. PSTT is diffusely and strongly positive, but choriocarcinoma and ETT are focally positive or negative, for hPL and Mel-CAM, which are markers of implantation site intermediate trophoblasts. However, as chorionic-type intermediate trophoblasts are positive for p63, the nucleus of ETT, which originates from them, is diffusely and strongly positive, but as PSTT is generally negative for p63, the staining behavior of PSTT and ETT to hPL/Mel-CAM and p63 is useful for their differentiation [7–9]. Recently, choriocarcinoma was reported to be positive, whereas PSTT and ETT are negative, for SALL4, which is a marker of germ cell tumors [10], and SALL4 can be a marker useful for the differentiation of the three diseases. The Ki-67 (MIB-1) index is high, being ≥50%, in choriocarcinoma, but is lower, by a few percent to 30%, in PSTT and ETT. The site of origin of ETT is the cervix to the lower body of the uterus in 30–50% of patients, and its differentiation from squamous cell carcinoma of the uterine cervix based on histomorphological

**Table 3.2** Immunohistochemical characteristics of choriocarcinoma, PSTT, and ETT

| Markers | Choriocarcinoma | PSTT | ETT |
|---|---|---|---|
| CK (CAM5.2) | + | + | + |
| hCG | ++ | + (focal)~ - | + (focal)~ – |
| hPL | ± | ++ | + (focal) ~ – |
| Mel-CAM (CD146) | + ~ ± | ++ | – ~ + (focal) |
| p63 | ± | – | ++ |
| SALL4 | + ~ ++ | – | – |
| PLAP | + | + | + ~ ± |
| Inhibin-α | + | + | + |
| Ki-67(MIB-1) index | >50% | 7~20% | 10~30% |

*CK* cytokeratin, *hCG* human chorionic gonadotropin, *hPL* human placental lactogen, *PLAP* placental alkaline phosphatase features is necessary. Both lesions are positive for p63, but although ETT is negative for p16, positive for inhibin-α, and positive for CK18, squamous cell carcinoma is generally positive for p16, negative for inhibin-α, and negative for CK18, making differentiation possible.

### *3.3.3 Treatments for PSTT and ETT*

PSTT and ETT are generally less responsive to chemotherapy than choriocarcinoma, and their basic treatment is surgery. For FIGO stage I, in which the lesion is localized in the uterus, total hysterectomy is recommended as the first choice [5, 6, 9]. In stage I, adjuvant therapy after total hysterectomy is considered unnecessary, but adjuvant chemotherapy using EP/EMA is recommended recently if there are any of the following risk factors: the period from antecedent (responsible) pregnancy to the onset is 2 years or longer, or there is a pathological finding of deep invasion, necrosis, or high mitotic rate (>5/10HPF) [5]. The survival rate of stage I patients is high at ≥90%. On the other hand, chemotherapy using EP/EMA in addition to maximum possible surgery is recommended for FIGO stage II or higher disease with metastasis or recurrence [5], but the survival rate of such cases is low, being 30–60%. Fertility preserving treatment for PSTT or ETT has not been established, and further accumulation of data is necessary.

## 3.4 Personalized Treatment for GTN—Future Potential and Perspectives

### *3.4.1 Anti-Angiogenic Therapy for GTN*

GTN usually has a rich blood supply. It has also been reported to frequently express vascular endothelial growth factor (VEGF) and its receptors [11]. However, there has been no clinical trial that demonstrated the efficacy of antiangiogenic agents for the treatment of GTN or evidence of the effectiveness of their combination with chemotherapy. There have been reports of the use of bevacizumab (anti-VEGF-A antibody) for refractory GTN, although they are few in number. Recently, Worley Jr. et al. reported a case of choriocarcinoma that resisted many kinds of chemotherapy regimens, but in which remission was achieved as the serum hCG level was normalized as a result of a marked response to two drug combination therapy using bevacizumab and anti-endoglin antibody (TRC105) [12]. Endoglin is a TGF-β1 co-receptor that plays an important role in vascular modeling. There are reports that endoglin is also expressed in human choriocarcinoma cells and that MTX, which is a key drug for choriocarcinoma, increases endoglin expression in choriocarcinoma cells [12, 13]. Indeed, endoglin was demonstrated to be expressed in tumor cells by immunohistochemical staining of lung metastases of choriocarcinoma, and the possibility that simultaneous inhibition of the VEGF and TGF-β/endoglin pathways is effective for the treatment of choriocarcinoma has been suggested [12]. From the above observations, VEGF and endoglin are expected to be potential targets of new targeted therapies against refractory GTN, and implementation of clinical trials and further basic studies is awaited.

### 3.4.2 Immune Tolerance Systems in GTN and Its Targeted Therapy

From the viewpoint that tumor cells of GTN originate from trophoblasts during pregnancy, GTN may be regarded as an "allograft or semi-allograft tumor" having paternal (partner) alloantigens. For GTN with such immunological characteristics to proliferate by escaping the attack of host immune cells, it is considered to have a powerful immune tolerance system similar to feto-maternal immune tolerance during pregnancy. Previous studies reported suppression of natural killer (NK) cells and activation of regulatory T cells (Treg) by the expression of human leukocyte antigen-G (HLA-G) in choriocarcinoma cells [14, 15] and immunosuppression via induction of Treg by hCG secreted from tumor cells and promotion of Th2 cytokine secretion [16]. In addition to these reports, it has been reported recently that there are immune tolerance systems in which immune checkpoint molecules play major roles in GTN as in many solid cancers. Programmed cell death 1 ligand 1 (PD-L1) is an immune checkpoint molecule expressed on the surface of tumor cells that binds to programmed cell death 1 (PD-1) on activated T cells and suppresses its function, and it is a major immune tolerance pathway of solid cancers [17]. In recent studies using immunohistochemistry, PD-L1 molecules were reported to be expressed in most tumor cells of GTN [18–21]. In contrast, indoleamine 2,3-dioxygenase (IDO), which is an enzyme that metabolizes tryptophan, strongly suppresses CD8 + cytotoxic T cells (CTL) and NK cells by depleting tryptophan and stimulating kynurenine production in the tumor microenvironment, is expressed in many solid cancers, and functions as an immune tolerance molecule [17, 22, 23]. According to our immunohistochemical evaluation of IDO expression in GTN, IDO was expressed in most tumor cells of choriocarcinoma and PSTT (unpublished data). Furthermore, interferon-γ (IFN-γ) secreted by CTL and NK promotes PD-L1 and IDO expression on the tumor side. For these reasons, PD-L1, IDO, and HLA-G are considered to be expressed in GTN and to be inducing tumor immune tolerance (Fig. 3.2).

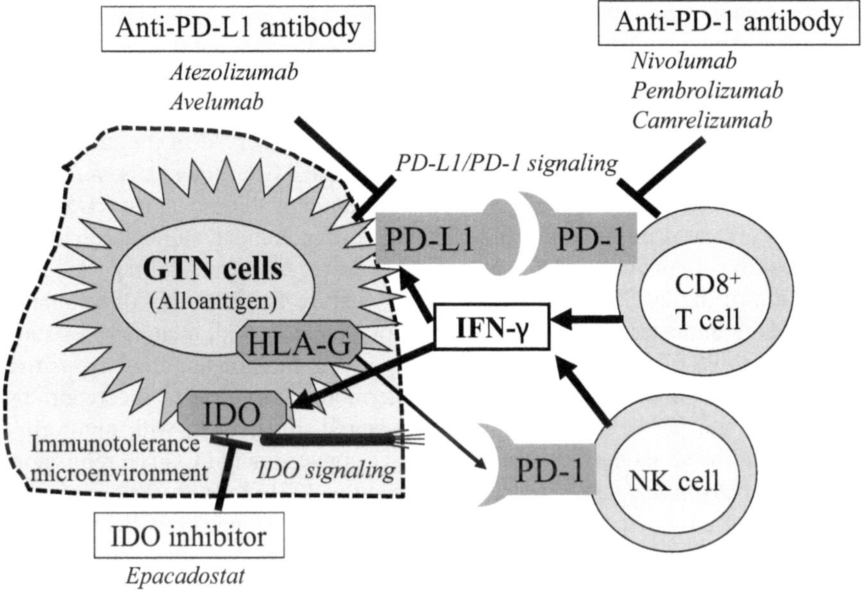


**Fig. 3.2** Immunotolerance-inducing molecules in GTN and their targeting therapy. When CTL and NK cells recognize and attack GTN with paternal alloantigens, interferon (IFN)-γ is locally produced. On the tumor cell side, the expression of PD-L1 or IDO, which are IFN-γ-dependent immunotolerant molecules, is considered to be induced, and they are considered to establish an immunotolerant condition by cooperating with HLA-G and other immunosuppressive factors. Immunotherapies using anti-PD-1 antibody, anti-PD-L1 antibody, or IDO inhibitor are expected to break through the immune tolerance and to be highly effective for the treatment of GTN. *IFN-γ* interferon-γ, *PD-L1* programmed cell death 1 ligand 1, *PD-1* programmed cell death 1, *NK* natural killer, *IDO* indoleamine 2,3-dioxygenase, *HLA-G* human leukocyte antigen-G

### 3.4.3 Immunotherapy for Chemo-Resistant GTN and Its Clinical Efficacy

Recently, a series of reports about the clinical efficacy of immune checkpoint inhibitors against refractory choriocarcinoma and PSTT/ETT based on evidence of induction of immunotolerance of GTN was published and is attracting attention. Reports published to the present are summarized in Table 3.3. Ghorani et al. administered pembrolizumab (anti-PD-1 antibody) to four patients with chemoresistant GTN (2 with choriocarcinoma, 1 with PSTT, and 1 with ETT) and reported that the treatment was ineffective in the ETT patient, but that the other three exhibited normalization of the hCG level by pembrolizumab single-drug therapy (4–8 cycles) and obtained durable complete response (CR) [24]. In this report, immunohistological evaluation was also performed, and in the patients with choriocarcinoma who obtained CR, tumor cells were positive for PD-L1 expression, CD8$^+$ tumor-infiltrating lymphocytes (TIL) were observed in large numbers, and many TIL were PD-1 positive. Tumor cells were also positive for HLA-G. In the ETT patient who was judged as having progressive disease (PD), tumor cells were PD-L1-positive, but no TIL accumulation was observed in the tumor, and HLA-G was negative. In the other patient who obtained CR, tumor cells were positive for PD-L1 and HLA-G, and a rich accumulation of PD-1-positive CD56$^+$NK cells was observed in the tumor [24]. Huang et al. administered pembrolizumab (2 cycles) to patients with treatment-resistant choriocarcinoma, and reported normalization of the hCG level and tumor cells positive for PD-L1 [25]. Similarly, Choi et al. administered pembrolizumab to two patients with treatment-resistant GTN (1 with PSTT and 1 with ETT), and reported remission and PD-L1-positive tumor cells by immunostaining [26]. Furthermore, according to the report by Goldfarb et al., pembrolizumab was administered to a patient with multidrug-resistant choriocarcinoma, resulting in negative conversion of hCG and remission, and tumor cells in this patient were PD-L1 positive [27]. Recently, You et al. carried out a phase II clinical trial of the PD-L1 inhibitor avelumab (TROPHIMMUN; NCT03135769) in 15 GTN patients with resistance to single-agent chemotherapy (12 with low-risk GTN and 3 with high-risk GTN), and reported that negative conversion of hCG and CR were obtained in 53.3% (8/15), that the profile of adverse events was safe, and remission was achieved by second-line chemotherapy in seven patients resistant to avelumab [28]. In this phase II trial, GTN patients resistant to multi-agent chemotherapy, including EMA/CO and EP/EMA, were also enrolled in Cohort B, and the results in this cohort are awaited. Lastly, Cheng et al. carried out a phase II trial (CAP01) of the combination of the PD-1 inhibitor camrelizumab and the VEGF receptor inhibitor apatinib in 20 GTN patients who were chemoresistant or had recurrence (19 with choriocarcinoma and 1 with PSTT), and reported that CR was obtained in 10 patients (50%) and that toxicity was within the acceptable range [29], suggesting the effectiveness of a combination of an immune checkpoint inhibitor and an anti-angiogenic agent. Although the number of reports is still small, the combination of PD-1 inhibitor/PD-L1 inhibitor may be effective for the treatment of refractory GTN, and PD-L1 and HLA-G expression in tumor cells and TIL accumulation in tumors may be biomarkers of responders to this treatment (Fig. 3.2). Moreover, recently, cases of refractory choriocarcinoma in which CR was obtained [30] and cases of ETT achieving PR [31] by treatment with pembrolizumab have been reported. Collectively, the CR achievement rate was high at nearly 50% or more, and the response rate of GTN to immune checkpoint inhibitors was higher than that of other solid cancers (Table 3.3). In the future, new personalized therapeutic strategies, such as combinations of these immune checkpoint inhibitors with the IDO inhibitor epacadostat [32], antiangiogenic agents, and existing chemotherapeutic agents, are expected to be developed. Theoretically, GTN with paternally derived alloantigens is a good target for immune checkpoint inhibitors, and implementation of further clinical trials of these immunotherapies for refractory chemoresistant cases and the establishment of their clinical use are urgently needed.

**Table 3.3** Reports on immunotherapy using immune checkpoint inhibitors for chemo-refractory gestational trophoblastic neoplasia (GTN)

| Author [Ref #] (Year) | Disease (Patient number) | Drug used | PD-L1 expression | Clinical outcome (CR%) |
|---|---|---|---|---|
| Ghorani [24] 2017 | Choriocarcinoma (*N* = 2)<br>PSTT (*N* = 1)<br>ETT (*N* = 1) | Pembrolizumab | All positive | CR3 (75%)<br>PD1 |
| Huang [25] 2017 | Choriocarcinoma (*N* = 1) | Pembrolizumab | Positive | CR (100%) |
| Choi [26] 2019 | PSTT (*N* = 1)<br>ETT (*N* = 1) | Pembrolizumab | Positive | CR2 (100%) |
| Goldfarb [27] 2020 | Choriocarcinoma (*N* = 1) | Pembrolizumab | Positive | CR (100%) |
| You [28] 2020 | Low-risk GTN (*N* = 12)<br>High-risk GTN (*N* = 3) | Avelumab | None | CR8 (53%) |
| Cheng [29] 2021 | Choriocarcinoma (*N* = 19)<br>PSTT (*N* = 1) | Camrelizumab (in combination with apatinib) | None | CR10 (50%) |
| Paspalj [30] 2021 | Choriocarcinoma (*N* = 1) | Pembrolizumab | None | CR (100%) |
| Bell [31] 2021 | ETT (*N* = 1) | Pembrolizumab | None | PR |

CR complete response, *PD* progressive disease, *PR* partial response, *PSTT* placental site trophoblastic tumor, *ETT* epithelioid trophoblastic tumor

## References


1. Ngan HYS, Seckl MJ, Berkowitz RS, Xiang Y, Golfier F, Sekharan PK, Lurain JR, Massuger L. Diagnosis and management of gestational trophoblastic disease: 2021 update. FIGO Cancer Report 2021. Int J Gynaecol Obstet. 2021;155(Suppl 1):86–93. https://doi.org/10.1002/ijgo.13877.
2. Lurain JR. Gestational trophoblastic disease I: epidemiology, pathology, clinical presentation and diagnosis of gestational trophoblastic disease, and management of hydatidiform mole. Am J Obstet Gynecol. 2010;203(6):531–9. https://doi.org/10.1016/j.ajog.2010.06.073. Epub 2010 Aug 21.PMID: 20728069 Review
3. Cheung AN, Hui P, Shih I, editors. 7. Gestational trophoblastic disease. Female genital Tumours. WHO classification of Tumours. 5th ed. Lyon: International Agency for Research on Cancer; 2020.
4. FIGO Oncology Committee. FIGO staging for gestational trophoblastic neoplasia 2000. Int J Gynecol Obstet. 2002;77:285–7.
5. NCCN Clinical Practice Guideline in Oncology. Gestational Trophoblastic Neoplasia, Version2. 2019, May 6., https://www.nccn.org/guidelines/guidelines-detail?category=1&id=1489
6. Seckl MJ, Sebire NJ, Fisher RA, et al. Gestational trophoblastic disease: ESMO clinical practice guidelines for diagnosis, treatment and follow-up. Ann Oncol. 2013;24:vi39–50.
7. Shih IM, et al. The pathology of intermediate trophoblastic tumors and tumor-like lesions. Int J Gynecol Pathol. 2001;20:31–47.
8. Shih IM, et al. Epithelioid trophoblastic tumor –a neoplasm distinct from choriocarcinoma and placental site trophoblastic tumor simulating carcinoma. Am J Surg Pathol. 1998;22:1393–403.

9. Horowitz NS, Goldstein DP, Berkowitz RS. Placental site trophoblastic tumors and epithelioid trophoblastic tumors: biology, natural history, and treatment modalities. Gynecol Oncol. 2017;144(1):208–14.
10. Stichelbout M, et al. SALL4 expression in gestational trophoblastic tumors: a useful tool to distinguish choriocarcinoma from placental site trophoblastic tumor and epithelioid trophoblastic tumor. Hum Pathol. 2016;54:121–6.
11. Singh M, Kindelberger D, Nagymanyoki Z, et al. Vascular endothelial growth factors and their receptors and regulators in gestational trophoblastic diseases and normal placenta. J Reprod Med. 2012;57:197–203.
12. Worley MJ Jr, Elias KM, Horowitz NS, Quade BJ, Berkowitz RS. Durable remission for a woman with refractory choriocarcinoma treated with anti-endoglin monoclonal antibody and bevacizumab: a case from the New England trophoblastic disease center, Brigham and Women's Hospital and Dana-Farber Cancer Institute. Gynecol Oncol. 2018;148(1):5–11.
13. Letamendía A, Lastres P, Almendro N, et al. Endoglin, a component of the TGF-beta receptor system, is a differentiation marker of human choriocarcinoma cells. Int J Cancer. 1998;76:541–6.
14. Wang X, Fu S, Freedman RS, et al. Immunobiology of gestational trophoblastic diseases. Int J Gynecol Cancer. 2006;16(4):1500–15.
15. Melsted WN, Matzen SH, Andersen MH, et al. The choriocarcinoma cell line JEG-3 upregulates regulatory T cell phenotypes and modulates pro-inflammatory cytokines through HLA-G. Cell Immunol. 2018;324:14–23.
16. Schumacher A. Human chorionic gonadotropin as a pivotal endocrine immune regulator initiating and preserving fetal tolerance. Int J Mol Sci. 2017;18(10):E2166.
17. Moon YW, Hajjar J, Hwu P, et al. Targeting the indoleamine 2,3-dioxygenase pathway in cancer. J Immunother Cancer. 2015;3:51.
18. Lu B, Teng X, Fu G, Bao L, Tang J, Shi H, Lu W, Lu Y. Analysis of PD-L1 expression in trophoblastic tissues and tumors. Hum Pathol. 2019 Feb;84:202–12.
19. Bolze PA, Patrier S, Massardier J, et al. PD-L1 expression in premalignant and malignant trophoblasts from gestational trophoblastic diseases is ubiquitous and independent of clinical outcomes. Int J Gynecol Cancer. 2017;27:554–61.
20. Veras E, Kurman RJ, Wang TL, et al. PD-L1 expression in human placentas and gestational trophoblastic diseases. Int J Gynecol Pathol. 2017;36:146–53.
21. Inaguma S, Wang Z, Lasota J, et al. Comprehensive immunohistochemical study of programmed cell death ligand 1 (PD-L1): analysis in 5536 cases revealed consistent expression in trophoblastic tumors. Am J Surg Pathol. 2016;40:1133–42.
22. Godin-Ethier J, Hanafi LA, Piccirillo CA, et al. Indoleamine 2,3-dioxygenase expression in human cancers: clinical and immunologic perspectives. Clin Cancer Res. 2011;17:6985–91.
23. Ino K. Indoleamine 2,3-dioxygenase and immune tolerance in ovarian cancer. Curr Opin Obstet Gynecol. 2011;23:13–8.
24. Ghorani E, Kaur B, Fisher RA, Short D, Joneborg U, Carlson JW, Akarca A, Marafioti T, Quezada SA, Sarwar N, Seckl MJ. Pembrolizumab is effective for drug-resistant gestational trophoblastic neoplasia. Lancet. 2017;390(10110):2343–5.
25. Huang M, Pinto A, Castillo RP, Slomovitz BM. Complete serologic response to pembrolizumab in a woman with chemoresistant metastatic choriocarcinoma. J Clin Oncol. 2017;35:3172–4.
26. Choi MC, Oh J, Lee C. Effective anti-programmed cell death 1 treatment for chemoresistant gestational trophoblastic neoplasia. Eur J Cancer. 2019;121:94–7.
27. Goldfarb JA, Dinoi G, Mariani A, Langstraat CL. A case of multi-agent drug resistant choriocarcinoma treated with Pembrolizumab. Gynecol Oncol Rep. 2020;32:100574.
28. You B, Bolze PA, Lotz JP, Massardier J, Gladieff L, Joly F, Hajri T, Maucort-Boulch D, Bin S, Rousset P, Devouassoux-Shisheboran M, Roux A, Alves-Ferreira M, Grazziotin-Soares D, Langlois-Jacques C, Mercier C, Villeneuve L, Freyer G, Golfier FJ. Avelumab in patients with gestational trophoblastic tumors with resistance to single-agent chemotherapy: cohort a of the TROPHIMMUN phase II trial. J Clin Oncol. 2020;38(27):3129–37.

29. Cheng H, Zong L, Kong Y, Wang X, Gu Y, Cang W, Zhao J, Wan X, Yang J, Xiang Y. Camrelizumab plus apatinib in patients with high-risk chemorefractory or relapsed gestational trophoblastic neoplasia (CAP 01): a single-arm, open-label, phase 2 trial. Lancet Oncol. 2021;22(11):1609–17.
30. Paspalj V, Polterauer S, Poetsch N, Reinthaller A, Grimm C, Bartl T. Long-term survival in multiresistant metastatic choriocarcinoma after pembrolizumab treatment: A case report Gynecol Oncol Rep. 2021;37:100817.
31. Bell SG, Uppal S, Sakala MD, Sciallis AP, Rolston A. An extrauterine extensively metastatic epithelioid trophoblastic tumor responsive to pembrolizumab. Gynecol Oncol Rep. 2021;37:100819.
32. Yao Y, Liang H, Fang X, Zhang S, Xing Z, Shi L, Kuang C, Seliger B, Yang Q. What is the prospect of indoleamine 2,3-dioxygenase 1 inhibition in cancer? Extrapolation from the past. J Exp Clin Cancer Res. 2021;40(1):60. https://doi.org/10.1186/s13046-021-01847-4.

# Chapter 4
# Personalized Treatment in Uterine Sarcoma

**Ken Yamaguchi**

**Abstract** Uterine sarcomas, including leiomyosarcoma and low- and high-grade endometrial stromal sarcoma, are rare but aggressive malignant diseases. Recent sequencing technology has identified pathogenic mutations and fusions of these malignancies. Leiomyosarcoma harbors frequent *TP53* and *RB1* mutations. Low-grade endometrial stromal sarcomas typically possess a *JAZF1-SUZ12* fusion, whereas high-grade endometrial stromal sarcomas harbor *YWHAE-NUTM2A/B* or *ZC3H7B-BCOR* fusions, or *BCOR* internal tandem duplications. Perivascular epithelioid cell tumors show inactivating mutations of *TSC1/TSC2* (which lead to activation of the mTOR pathway) or *TFE3* fusions. Fibrosarcoma-like tumors originating in the cervix and/or lower uterine segment sometimes harbors NTRK gene rearrangements or *COL1A1-PDGFB* translocation. Somatic *DICER1* mutations are identified in a small subset of embryonal rhabdomyosarcoma of the cervix. Molecular targeting strategies based on these genomic alterations are needed to treat uterine sarcomas.

**Keywords** Leiomyosarcoma · Endometrial stromal sarcoma · Undifferentiated uterine sarcoma · Perivascular epithelioid cell tumor: PEComa · Adenosarcoma · Rhabdomyosarcoma · Mutation · Fusion gene

## 4.1 Introduction

Gynecologic sarcomas represent 3–4% of all gynecologic malignancies and 13% of all mesenchymal malignancies [1]. The uterus is the most frequent primary site (83%), whereas the ovary (8%), vulva and vagina (5%), and other gynecologic organs (2%) are less frequently involved. Gynecologic sarcomas, especially uterine sarcomas, exhibit extremely aggressive behaviors with poor prognosis and a high mortality rate. Benign

K. Yamaguchi (✉)
Department of Gynecology and Obstetrics, Kyoto University Graduate School of Medicine, Kyoto, Japan
e-mail: soulken@kuhp.kyoto-u.ac.jp


M. Mandai (ed.), *Personalization in Gynecologic Oncology*, Comprehensive Gynecology and Obstetrics, https://doi.org/10.1007/978-981-19-4711-7_4

mesenchymal tumors and those with uncertain malignant potential also affect the gynecological tract and have similar morphology with more aggressive sarcomas. In recent years, the discovery of genetic abnormalities and molecular alterations has led to a deeper understanding of gynecologic sarcomas, indicating refined classification systems and individualized therapies. A dynamic evolution from "morphological classification" to "molecular classification" has occurred in this field of uterine sarcoma.

In this chapter, "molecular classification" and "personalized therapies" in uterine sarcoma will be described according to the "morphological classification" of uterine sarcoma.

## 4.2 Leiomyosarcoma

Uterine leiomyosarcoma is a malignant mesenchymal tumor of myometrial smooth muscle cells. Leiomyosarcomas are the most common type of uterine sarcoma (30–50%) but only account for 1–2% of all uterine malignancies [2]. These tumor cells express desmin, h-caldexmon, and SMA. When the tumor is a poorly differentiated or myxoid subtype, these expressions may be weak and/or patchy [3, 4]. Positivity for CD10, EMA, and cytokeratin is common in epithelioid tumors [5, 6]. ER, PR, p16, and p53 also show positive expression [4, 5].

The Stanford criteria are used to classify uterine leiomyosarcoma based on the frequency of atypical mitosis, cytological atypia, and tumor cell necrosis. Because the Stanford criteria are composed of factors related to morphological aggressiveness, the criteria are associated with the prognosis and aggressive behaviors. Ki67 is the standard for the evaluation of the proliferative activity of tumors. Ki67 labeling index ≥10% was reported as a prognostic factor in uterine leiomyosarcoma [7].

Various molecular abnormalities have been identified in uterine leiomyosarcomas, but none are considered definitive pathogenesis. *TP53* mutations are identified in 30–60% of uterine leiomyosarcomas, followed by *RB1* (50%), *ATX* (−30%), *MED12* (−20%), *PTEN* (−20%), and *BRCA2* (8%) mutations [8–10]. Clinical benefit was reported in patients with uterine leiomyosarcoma with somatic *BRCA2* alterations treated with PARP inhibitors [8].

Hormonal therapies are alternative strategies, especially for advanced and recurrent diseases. In uterine leiomyosarcoma, ER and/or PR expression is positive in approximately 40–80% [1]. Small retrospective studies have reported a limited efficacy of aromatase inhibitors in patients with advanced and metastatic uterine leiomyosarcoma. Although the response rate is low, a relatively prolonged progression-free survival has been observed.

## 4.3 Low-Grade Endometrial Stromal Sarcoma

Low-grade endometrial stromal sarcoma is a malignant stromal tumor with cells resembling proliferative-phase endometrial stroma and displaying infiltrative growth. Immunohistochemically, low-grade endometrial stromal sarcoma usually shows a diffuse expression of CD10, IFITM1, ER, and PR, with focal cyclin D1 positivity [5, 11–15]. Tumors may be positive for cytokeratins [16], muscle markers (desmin, SMA, and h-caldesmon in areas of smooth muscle differentiation), and are often positive in sex cord-like elements, which also express inhibin, calretinin, melan-A, WT1, and CD99 [5, 17–21].

Two-thirds of low-grade endometrial stromal sarcoma harbor genetic fusions involving polycomb family genes, with *JAZF1-SUZ12* being most common, followed by *JAZF1-JJAZ1*, *JAZF1-PHF1*, *EPC1-PHF1*, *MEAF6-PHF1*, *MBTD1-EZHIP*, *BRD8-PHF1*, *EPC2-PHF1*, *EPC1-SUZ12*, and *ZC3H7-BCOR* [17, 22–42]. The presence or absence of known detectable translocations is not associated with distinct clinical behavior in low-grade endometrial stromal sarcoma [43]. Case series identified that the *ESR1* p.Y537S hotspot mutation in low-grade endometrial stromal sarcoma with high-grade histologic transformation might be associated with endocrine resistance in these lesions [44]. The molecular classifications of low-grade endometrial stromal sarcoma are listed in Table 4.1.

**Table 4.1** Molecular classification of low-grade endometrial stromal sarcoma

| Gene rearrangement | Fusion gene | Protein function | Diagnostic method |
|---|---|---|---|
| t(7;17)(p15;q21) | *JAZF1-SUZ12* | JAZF1: A nuclear protein with zinc fingers and functions as a transcriptional repressor.<br>SUZ12: Polycomb protein and functions with chromatin silencing using its zinc-finger domain to bind RNA. | FISH, PCR |
| t(6;7)(p21;p15) | *JAZF1-PHF1* | JAZF1: A nuclear protein with zinc fingers and functions as a transcriptional repressor.<br>PHF1: Polycomb group protein, a component of a histone H3 lysine-27 (H3K27)-specific methyltransferase complex, and functions with transcriptional regulation by influencing chromatin structure. | FISH |
| t(6;10)(p21;p11) | *EPC1-PHF1* | EPC1: Polycomb protein, a component of the NuA4 histone acetyltransferase complex, and functions with a transcriptional activator and repressor.<br>PHF1: Polycomb group protein, a component of a histone H3 lysine-27 (H3K27)-specific methyltransferase complex, and functions with transcriptional regulation by influencing chromatin structure. | FISH |
| t(1;6)(p34;p21) | *MEAF6-PHF1* | MEAF6: A component of the NuA4 histone acetyltransferase complex and functions with a transcriptional activator and repressor.<br>PHF1: Polycomb group protein, a component of a histone H3 lysine-27 (H3K27)-specific methyltransferase complex, and functions with transcriptional regulation by influencing chromatin structure. | FISH |
| t(X;17)(p11;q21) | *CXorf67-MBTD1* | CXorf67: A repressor of PRC2/EED-EZH1 and PRC2/EED-EZH2 complex function by inhibiting EZH1/EZH2 methyltransferase activity.<br>MBTD1: Putative polycomb group protein and functions with the maintenance of the transcriptionally repressive state of genes | FISH, PCR |

(continued)

**Table 4.1** (continued)

| Gene rearrangement | Fusion gene | Protein function | Diagnostic method |
|---|---|---|---|
| t(X;22) (p11;q13) | *ZC3H7B-BCOR* | ZC3H7B: A possible regulator of miRNA biogenesis BCOR: An interacting corepressor of BCL6, a POZ/zinc finger transcription repressor, and functions by interacting with class I and II histone deacetylases. | FISH, PCR |

## 4.4 High-Grade Endometrial Stromal Sarcoma

High-grade endometrial stromal sarcoma is a malignant endometrial stromal tumor with uniform high-grade round and/or spindle morphology, sometimes with a low-grade component. The WHO classifications have been changed according to genetic backgrounds. High-grade endometrial stromal sarcoma and undifferentiated uterine sarcoma were categorized together in the 2003 World Health Organization (WHO) classification. In the 2014 WHO classification, however, high-grade endometrial stromal sarcoma was categorized independently from undifferentiated uterine sarcoma because *YWHAE-NUTM2A/B* (previously referred to as *YWHAE-FAM22A/B*) fusion genes were identified in high-grade endometrial stromal sarcoma [45]. *BCOR* internal tandem duplications (ITD) were identified as an oncogenic alternative to *YWHAE-NUTM2* fusion in high-grade uterine sarcomas [46]. Therefore, demonstration of gene fusion or ITD may be useful for diagnosing high-grade endometrial stromal sarcoma. The high-grade component of this disease with *YWHAE-NUTM2A/B* fusion is positive for cyclin D1, BCOR, KIT, CD56, and CD99; negative for CD10 and DOG1; and either negative or weakly positive for ER and PR [47–49]. No mutations of the *KIT* gene have been detected [50]. The morphologically low-grade spindle cells are positive for CD10, ER, and PR and negative for cyclinD1; BCOR is variable.

Other pathogenetic gene alterations include *ZC3H7B-BCOR* fusions [42, 51–53], or *BCOR* ITD [46, 49, 54], *EPC1-BCOR*, *JAZF1-BCORL1*, and *BRD8-PHF1* fusions [41, 42, 55]. *ZC3H7B-BCOR* high-grade endometrial stromal sarcomas are typically positive for cyclin D1, whereas only about 50% of the cases express BCOR. CD10 with variable ER and PR positivity shows diffuse positive in the typical cases. SMA and caldesmon show focal positive expression, but the expression of desmin is negative. Although the tumors show pan-TRK staining, it is not related to NTRK rearrangement [56], suggesting the limited efficacy of TRK inhibitors. *BCOR* ITD tumors have a distinct immunoprofile from those of *ZC3H7B-BCO* tumors: expression of CD10 is less positive, cyclin D1 and BCOR are diffuse positive, and ER and PR are negative. They are positive for desmin but are negative for SMA and caldesmon [46, 49, 57]. Positive immunohistochemistry for cyclin D1 and positive immunohistochemistry for *ZC3H7B-BCOR*, *YWHAE-NUTM2A/B* (*FAM22A/B*), or *BCOR* ITD should be essential for the diagnosis of high-grade endometrial stromal sarcoma. Table 4.2 summarizes the immunophenotype of uterine mesenchymal tumors, including endometrial stromal tumors.

**Table 4.2** Biomarkers of uterine mesenchymal tumors

| | ER/PR | CD10 | IFITM1 | CDK1 | BCOR | cKit | Keratin | SMA | Desmin | Caldesmon | Calponin |
|---|---|---|---|---|---|---|---|---|---|---|---|
| Endometrial stromal nodule | +/+ | + | | Focal | Focal | | | + | −/+ | + | |
| Low-grade endometrial stromal sarcoma | +/+ | + | + | Focal | Focal | − | + | + | −/+ | + | + |
| *YWHAE-NUTM2* (low-grade component) | +/+ | + | | Focal | Focal | | − | − | − | − | |
| *YWHAE-NUTM2* (high-grade component) | −/− | − | + | + | + | + | − | − | − | − | |
| *ZC3H7B-BCOR* | +/+ | + | | + | + | + | − | Focal | − | Focal | |
| *BCOR* ITD | −/− | + | | + | + | | | − | Focal | − | |
| Undifferentiated uterine sarcoma | −/− | Variable | | Variable | − | − | Focal | Focal | Focal | | |
| Leiomyosarcoma | Variable | −/+ | | | | Variable | + | + | + | + | + |

## 4.5 Undifferentiated Uterine Sarcoma

In the 2014 WHO classification, undifferentiated uterine sarcoma was again recategorized independently from high-grade endometrial stromal sarcoma. Undifferentiated uterine sarcoma is a malignant mesenchymal tumor that originates in the uterus, lacking evidence of specific lines of differentiation. Therefore, this is a heterogeneous group and a diagnosis of exclusion. *YWHAE*, *JAZF1*, and *NTRK* rearrangements should not be classified in the category. Undifferentiated uterine sarcomas are positive for p53 and p16, and some of them are positive for ER and/or PR and variable positivity for CD10 [58]. Some undifferentiated uterine sarcomas are associated with a low-grade endometrial stromal component and diffusely express cyclin D1. Positive immunohistochemistry for *ZC3H7B-BCOR*, *YWHAE-NUTM2A/B* (*FAM22A/B*), and *BCOR* ITD should be classified in high-grade endometrial stromal sarcoma.

## 4.6 Perivascular Epithelioid Cell Tumor: PEComa

Perivascular epithelioid cell tumors (PEComas) are mesenchymal neoplasms composed of perivascular epithelioid cells that express melanocytic and smooth muscle markers. Although most PEComas are benign with no recurrence potential after complete surgical excision, some of them exhibit malignant behavior. The criteria of malignant PEComa are proposed by Schoolmeester et al. [59]. The presence of more than four features (gross size ≥5 cm, high-grade nuclear features, necrosis, vascular invasion, or a mitotic rate ≥ 1/50 HPF) indicates malignant PEComas. Inactivating mutations of *TSC1/TSC2* are observed in PEComa. Some have *TFE3*, *RAD51B*, or *HTR4-ST3GAL1* fusions [60–63]. TSC mutations and TFE3 fusions are mutually exclusive [64]. Inactivating mutations of TSC1/TSC2 activate the mTOR pathway [65], indicating that mTOR inhibitors are a rational targeting therapy in PEComa [66]. Conventional PEComas exhibit high sensitivity for HMB45 and are specific for melan-A. PEComas express SMA, dsmin, and h-caldesmon. Cathepsin K is positive in essentially all tumors, whereas only a minority of PEComa patients express S100, MITF, and tyrosinase. *TFE3* translocation-associated PEComas show diffuse positive for TFE3, HMB34, and cathepsin K, with focal to absent melan-A.

## 4.7 Adenosarcoma

Adenosarcoma is a biphasic neoplasm comprising a benign epithelial component and a malignant stromal component and, in the 2014 WHO classification, is categorized as a mixed epithelial and mesenchymal tumor. When the sarcoma component is more than 25%, it is defined as sarcomatous overgrowth, which is transformed into high-grade sarcoma. Immunohistochemistry exhibits positive for CD10, ER, and PR, although these are often negative in sarcomatous overgrowth [67, 68]. There are definitive pathogenic molecular alterations. However, mutations of the PIK3/AKT/PTEN pathway are described in 70% of adenosarcomas [69]. Although *TP53* mutation is uncommon, *TP53* mutation is associated with sarcomatous overgrowth and aggressive behavior [70–72].

## 4.8 NTRK-Rearranged Spindle Cell Neoplasm Resembling Fibrosarcoma

Neoplasm originating in the cervix and/or lower uterine segment with fibrosarcoma-like morphology often harbors NTRK gene rearrangements. Patients with NTRK-rearranged cervical sarcoma range in age from 20 to 45 years [73, 74]. The NTRK-rearranged spindle cells express S100, CD34, TRK, and cyclin D1. No expression of CD10, SMA, desmin, BCOR, ER, or PR is observed [73, 75]. NTRK inhibitors have shown dramatic and durable responses in advanced NTRK-rearranged sarcoma cases.

## 4.9 COL1A1-PDGFB Translocation-Associated Fibrosarcoma

Uterine and vaginal sarcomas resembling fibrosarcoma sometimes harbor the *COL1A1-PDGFB* fusion. Uterine and vaginal sarcomas with *COL1A1-PDGFB* translocation were seen in older patients (more than 45 years old) [75]. They strongly express CD34 but not S100, Trk, ER, PR, or desmin [75]. Imatinib (tyrosine kinase inhibitor) is a promising agent against *COL1A1-PDGFB* translocation-associated sarcoma.

## 4.10 Rhabdomyosarcoma

Rhabdomyosarcomas are malignant mesenchymal tumors exhibiting skeletal/muscle differentiation. Three subtypes are listed in the 2014 WHO classification; pleomorphic rhabdomyosarcoma, embryonal rhabdomyosarcoma, and alveolar rhabdomyosarcoma. Pleomorphic rhabdomyosarcoma is typically originated in the uterine corpus. Embryonal rhabdomyosarcoma is most common in the vagina in children and the cervix and corpus in adolescents and adults. Alveolar rhabdomyosarcoma commonly occurs in the vulva [76–80]. Somatic *DICER1* mutations have been reported in a small subset of embryonal rhabdomyosarcoma of the cervix [81, 82]. In alveolar rhabdomyosarcoma, *PAX3-FOXO1* or *PAX7-FOXO1* fusion genes encoding transcript factors that activate *PAX3/PAX7* target genes are occasionally identified. Among metastatic childhood alveolar rhabdomyosarcoma, PAX3-FKHR-positive patients exhibited a poorer prognosis than PAX7-FKHR-positive patients [83]. Rhabdomyosarcomas are positive for myogenin and MYOD1, which is more diffuse in alveolar rhabdomyosarcoma than embryonal rhabdomyosarcoma [84]. MSA, desmin, myoglobin, and myosin may be positive, but negative for SMA.

## 4.11 Conclusion

Sequencing technology allows us to categorize uterine mesenchymal malignant tumors based on the pathogenic molecular alterations. Several mesenchymal tumors may be treatable using therapies focused on promising targets: mTOR inhibitors for PEComa, NTRK inhibitors for NTRK-rearranged sarcoma, and Imatinib (tyrosine kinase inhibitor) for *COL1A1-PDGFB* translocation-associated sarcoma. Further novel targeted therapies should be developed based on the molecular pathogenesis of the aggressive disease uterine sarcoma.

## References

1. Libertini M, Hallin M, Thway K, Noujaim J, Benson C, van der Graaf W, et al. Gynecological sarcomas: molecular characteristics, behavior, and histology-driven therapy. Int J Surg Pathol. 2021;29(1):4–20.
2. Lin JF, Slomovitz BM. Uterine sarcoma 2008. Curr Oncol Rep. 2008;10(6):512–8.
3. Mills AM, Longacre TA. Smooth muscle tumors of the female genital tract. Surg Pathol Clin. 2009;2(4):625–77.
4. Busca A, Parra-Herran C. Myxoid mesenchymal tumors of the uterus: An update on classification, definitions, and differential diagnosis. Adv Anat Pathol. 2017;24(6):354–61.
5. Oliva E, Young RH, Amin MB, Clement PB. An immunohistochemical analysis of endometrial stromal and smooth muscle tumors of the uterus: a study of 54 cases emphasizing the importance of using a panel because of overlap in immunoreactivity for individual antibodies. Am J Surg Pathol. 2002;26(4):403–12.
6. Iwata J, Fletcher CD. Immunohistochemical detection of cytokeratin and epithelial membrane antigen in leiomyosarcoma: a systematic study of 100 cases. Pathol Int. 2000;50(1):7–14.
7. Travaglino A, Raffone A, Catena U, De Luca M, Toscano P, Del Prete E, et al. Ki67 as a prognostic marker in uterine leiomyosarcoma: a quantitative systematic review. Eur J Obstet Gynecol Reprod Biol. 2021;266:119–24.
8. Hensley ML, Chavan SS, Solit DB, Murali R, Soslow R, Chiang S, et al. Genomic landscape of uterine sarcomas defined through prospective clinical sequencing. Clin Cancer Res. 2020;26(14):3881–8.
9. An Y, Wang S, Li S, Zhang L, Wang D, Wang H, et al. Distinct molecular subtypes of uterine leiomyosarcoma respond differently to chemotherapy treatment. BMC Cancer. 2017;17(1):639.

10. Yang CY, Liau JY, Huang WJ, Chang YT, Chang MC, Lee JC, et al. Targeted next-generation sequencing of cancer genes identified frequent TP53 and ATRX mutations in leiomyosarcoma. Am J Transl Res. 2015;7(10):2072–81.
11. Chu PG, Arber DA, Weiss LM, Chang KL. Utility of CD10 in distinguishing between endometrial stromal sarcoma and uterine smooth muscle tumors: an immunohistochemical comparison of 34 cases. Mod Pathol. 2001;14(5):465–71.
12. McCluggage WG, Sumathi VP, Maxwell P. CD10 is a sensitive and diagnostically useful immunohistochemical marker of normal endometrial stroma and of endometrial stromal neoplasms. Histopathology. 2001;39(3):273–8.
13. Reich O, Regauer S, Urdl W, Lahousen M, Winter R. Expression of oestrogen and progesterone receptors in low-grade endometrial stromal sarcomas. Br J Cancer. 2000;82(5):1030–4.
14. Lee CH, Ali RH, Rouzbahman M, Marino-Enriquez A, Zhu M, Guo X, et al. Cyclin D1 as a diagnostic immunomarker for endometrial stromal sarcoma with YWHAE-FAM22 rearrangement. Am J Surg Pathol. 2012;36(10):1562–70.
15. Busca A, Gulavita P, Parra-Herran C, Islam S. IFITM1 outperforms CD10 in differentiating low-grade endometrial stromal sarcomas from smooth muscle neoplasms of the uterus. Int J Gynecol Pathol. 2018;37(4):372–8.
16. Farhood AI, Abrams J. Immunohistochemistry of endometrial stromal sarcoma. Hum Pathol. 1991;22(3):224–30.
17. Baker P, Oliva E. Endometrial stromal tumours of the uterus: a practical approach using conventional morphology and ancillary techniques. J Clin Pathol. 2007;60(3):235–43.
18. Baker RJ, Hildebrandt RH, Rouse RV, Hendrickson MR, Longacre TA. Inhibin and CD99 (MIC2) expression in uterine stromal neoplasms with sex-cord-like elements. Hum Pathol. 1999;30(6):671–9.
19. Irving JA, Carinelli S, Prat J. Uterine tumors resembling ovarian sex cord tumors are polyphenotypic neoplasms with true sex cord differentiation. Mod Pathol. 2006;19(1):17–24.
20. Moinfar F, Regitnig P, Tabrizi AD, Denk H, Tavassoli FA. Expression of androgen receptors in benign and malignant endometrial stromal neoplasms. Virchows Arch. 2004;444(5):410–4.
21. Sumathi VP, Al-Hussaini M, Connolly LE, Fullerton L, McCluggage WG. Endometrial stromal neoplasms are immunoreactive with WT-1 antibody. Int J Gynecol Pathol. 2004;23(3):241–7.
22. Chiang S, Ali R, Melnyk N, McAlpine JN, Huntsman DG, Gilks CB, et al. Frequency of known gene rearrangements in endometrial stromal tumors. Am J Surg Pathol. 2011;35(9):1364–72.
23. Koontz JI, Soreng AL, Nucci M, Kuo FC, Pauwels P, van Den Berghe H, et al. Frequent fusion of the JAZF1 and JJAZ1 genes in endometrial stromal tumors. Proc Natl Acad Sci U S A. 2001;98(11):6348–53.
24. Micci F, Walter CU, Teixeira MR, Panagopoulos I, Bjerkehagen B, Saeter G, et al. Cytogenetic and molecular genetic analyses of endometrial stromal sarcoma: nonrandom involvement of chromosome arms 6p and 7p and confirmation of JAZF1/JJAZ1 gene fusion in t(7;17). Cancer Genet Cytogenet. 2003;144(2):119–24.
25. Huang HY, Ladanyi M, Soslow RA. Molecular detection of JAZF1-JJAZ1 gene fusion in endometrial stromal neoplasms with classic and variant histology: evidence for genetic heterogeneity. Am J Surg Pathol. 2004;28(2):224–32.
26. Hrzenjak A, Moinfar F, Tavassoli FA, Strohmeier B, Kremser ML, Zatloukal K, et al. JAZF1/JJAZ1 gene fusion in endometrial stromal sarcomas: molecular analysis by reverse transcriptase-polymerase chain reaction optimized for paraffin-embedded tissue. J Mol Diagn. 2005;7(3):388–95.
27. Micci F, Panagopoulos I, Bjerkehagen B, Heim S. Consistent rearrangement of chromosomal band 6p21 with generation of fusion genes JAZF1/PHF1 and EPC1/PHF1 in endometrial stromal sarcoma. Cancer Res. 2006;66(1):107–12.
28. Nara M, Hashi A, Murata S, Kondo T, Yuminamochi T, Nakazawa K, et al. Lobular endocervical glandular hyperplasia as a presumed precursor of cervical adenocarcinoma independent of human papillomavirus infection. Gynecol Oncol. 2007;106(2):289–98.

29. Kurihara S, Oda Y, Ohishi Y, Iwasa A, Takahira T, Kaneki E, et al. Endometrial stromal sarcomas and related high-grade sarcomas: immunohistochemical and molecular genetic study of 31 cases. Am J Surg Pathol. 2008;32(8):1228–38.
30. Jakate K, Azimi F, Ali RH, Lee CH, Clarke BA, Rasty G, et al. Endometrial sarcomas: an immunohistochemical and JAZF1 re-arrangement study in low-grade and undifferentiated tumors. Mod Pathol. 2013;26(1):95–105.
31. D'Angelo E, Ali RH, Espinosa I, Lee CH, Huntsman DG, Gilks B, et al. Endometrial stromal sarcomas with sex cord differentiation are associated with PHF1 rearrangement. Am J Surg Pathol. 2013;37(4):514–21.
32. Stewart CJ, Leung YC, Murch A, Peverall J. Evaluation of fluorescence in-situ hybridization in monomorphic endometrial stromal neoplasms and their histological mimics: a review of 49 cases. Histopathology. 2014;65(4):473–82.
33. Ali RH, Al-Safi R, Al-Waheeb S, John B, Al-Ali W, Al-Jassar W, et al. Molecular characterization of a population-based series of endometrial stromal sarcomas in Kuwait. Hum Pathol. 2014;45(12):2453–62.
34. Hodge JC, Bedroske PP, Pearce KE, Sukov WR. Molecular cytogenetic analysis of JAZF1, PHF1, and YWHAE in endometrial stromal tumors: discovery of genetic complexity by fluorescence in situ hybridization. J Mol Diagn. 2016;18(4):516–26.
35. Micci F, Gorunova L, Agostini A, Johannessen LE, Brunetti M, Davidson B, et al. Cytogenetic and molecular profile of endometrial stromal sarcoma. Genes Chromosomes Cancer. 2016;55(11):834–46.
36. Panagopoulos I, Micci F, Thorsen J, Gorunova L, Eibak AM, Bjerkehagen B, et al. Novel fusion of MYST/Esa1-associated factor 6 and PHF1 in endometrial stromal sarcoma. PLoS One. 2012;7(6):e39354.
37. Micci F, Gorunova L, Gatius S, Matias-Guiu X, Davidson B, Heim S, et al. MEAF6/PHF1 is a recurrent gene fusion in endometrial stromal sarcoma. Cancer Lett. 2014;347(1):75–8.
38. Dewaele B, Przybyl J, Quattrone A, Finalet Ferreiro J, Vanspauwen V, Geerdens E, et al. Identification of a novel, recurrent MBTD1-CXorf67 fusion in low-grade endometrial stromal sarcoma. Int J Cancer. 2014;134(5):1112–22.
39. Micci F, Brunetti M, Dal Cin P, Nucci MR, Gorunova L, Heim S, et al. Fusion of the genes BRD8 and PHF1 in endometrial stromal sarcoma. Genes Chromosomes Cancer. 2017;56(12):841–5.
40. Brunetti M, Gorunova L, Davidson B, Heim S, Panagopoulos I, Micci F. Identification of an. Oncotarget. 2018;9(27):19203–8.
41. Dickson BC, Lum A, Swanson D, Bernardini MQ, Colgan TJ, Shaw PA, et al. Novel EPC1 gene fusions in endometrial stromal sarcoma. Genes Chromosomes Cancer. 2018;57(11):598–603.
42. Cotzia P, Benayed R, Mullaney K, Oliva E, Felix A, Ferreira J, et al. Undifferentiated uterine sarcomas represent under-recognized high-grade endometrial stromal sarcomas. Am J Surg Pathol. 2019;43(5):662–9.
43. Lee CH, Nucci MR. Endometrial stromal sarcoma--the new genetic paradigm. Histopathology. 2015;67(1):1–19.
44. Dessources K, Miller KM, Kertowidjojo E, Da Cruz PA, Zou Y, Selenica P, et al. ESR1 hotspot mutations in endometrial stromal sarcoma with high-grade transformation and endocrine treatment. Mod Pathol. 2021;35(7):972–8.
45. Lee CH, Ou WB, Mariño-Enriquez A, Zhu M, Mayeda M, Wang Y, et al. 14-3-3 fusion oncogenes in high-grade endometrial stromal sarcoma. Proc Natl Acad Sci U S A. 2012;109(3):929–34.
46. Mariño-Enriquez A, Lauria A, Przybyl J, Ng TL, Kowalewska M, Debiec-Rychter M, et al. BCOR internal tandem duplication in high-grade uterine sarcomas. Am J Surg Pathol. 2018;42(3):335–41.
47. Lee CH, Hoang LN, Yip S, Reyes C, Marino-Enriquez A, Eilers G, et al. Frequent expression of KIT in endometrial stromal sarcoma with YWHAE genetic rearrangement. Mod Pathol. 2014;27(5):751–7.
48. Lee CH, Mariño-Enriquez A, Ou W, Zhu M, Ali RH, Chiang S, et al. The clinicopathologic features of YWHAE-FAM22 endometrial stromal sarcomas: a histologically high-grade and clinically aggressive tumor. Am J Surg Pathol. 2012;36(5):641–53.

49. Chiang S, Lee CH, Stewart CJR, Oliva E, Hoang LN, Ali RH, et al. BCOR is a robust diagnostic immunohistochemical marker of genetically diverse high-grade endometrial stromal sarcoma, including tumors exhibiting variant morphology. Mod Pathol. 2017;30(9):1251–61.
50. Sciallis AP, Bedroske PP, Schoolmeester JK, Sukov WR, Keeney GL, Hodge JC, et al. High-grade endometrial stromal sarcomas: a clinicopathologic study of a group of tumors with heterogenous morphologic and genetic features. Am J Surg Pathol. 2014;38(9):1161–72.
51. Panagopoulos I, Thorsen J, Gorunova L, Haugom L, Bjerkehagen B, Davidson B, et al. Fusion of the ZC3H7B and BCOR genes in endometrial stromal sarcomas carrying an X;22-translocation. Genes Chromosomes Cancer. 2013;52(7):610–8.
52. Lewis N, Soslow RA, Delair DF, Park KJ, Murali R, Hollmann TJ, et al. ZC3H7B-BCOR high-grade endometrial stromal sarcomas: a report of 17 cases of a newly defined entity. Mod Pathol. 2018;31(4):674–84.
53. Hoang LN, Aneja A, Conlon N, Delair DF, Middha S, Benayed R, et al. Novel high-grade endometrial stromal sarcoma: a morphologic mimicker of Myxoid Leiomyosarcoma. Am J Surg Pathol. 2017;41(1):12–24.
54. Juckett LT, Lin DI, Madison R, Ross JS, Schrock AB, Ali S. A Pan-cancer landscape analysis reveals a subset of endometrial stromal and pediatric tumors defined by internal tandem duplications of BCOR. Oncology. 2019;96(2):101–9.
55. Allen AJ, Ali SM, Gowen K, Elvin JA, Pejovic T. A recurrent endometrial stromal sarcoma harbors the novel fusion. Gynecol Oncol Rep. 2017;20:51–3.
56. Solomon JP, Linkov I, Rosado A, Mullaney K, Rosen EY, Frosina D, et al. NTRK fusion detection across multiple assays and 33,997 cases: diagnostic implications and pitfalls. Mod Pathol. 2020;33(1):38–46.
57. Li W, Li L, Wu M, Lang J, Bi Y. The prognosis of stage IA mixed endometrial carcinoma. Am J Clin Pathol. 2019;152(5):616–24.
58. Halbwedl I, Ullmann R, Kremser ML, Man YG, Isadi-Moud N, Lax S, et al. Chromosomal alterations in low-grade endometrial stromal sarcoma and undifferentiated endometrial sarcoma as detected by comparative genomic hybridization. Gynecol Oncol. 2005;97(2):582–7.
59. Schoolmeester JK, Howitt BE, Hirsch MS, Dal Cin P, Quade BJ, Nucci MR. Perivascular epithelioid cell neoplasm (PEComa) of the gynecologic tract: clinicopathologic and immunohistochemical characterization of 16 cases. Am J Surg Pathol. 2014;38(2):176–88.
60. Argani P, Aulmann S, Illei PB, Netto GJ, Ro J, Cho HY, et al. A distinctive subset of PEComas harbors TFE3 gene fusions. Am J Surg Pathol. 2010;34(10):1395–406.
61. Schoolmeester JK, Dao LN, Sukov WR, Wang L, Park KJ, Murali R, et al. TFE3 translocation-associated perivascular epithelioid cell neoplasm (PEComa) of the gynecologic tract: morphology, immunophenotype, differential diagnosis. Am J Surg Pathol. 2015;39(3):394–404.
62. Bennett JA, Braga AC, Pinto A, Van de Vijver K, Cornejo K, Pesci A, et al. Uterine PEComas: a morphologic, Immunohistochemical, and molecular analysis of 32 tumors. Am J Surg Pathol. 2018;42(10):1370–83.
63. Agaram NP, Sung YS, Zhang L, Chen CL, Chen HW, Singer S, et al. Dichotomy of genetic abnormalities in PEComas with therapeutic implications. Am J Surg Pathol. 2015;39(6):813–25.
64. Malinowska I, Kwiatkowski DJ, Weiss S, Martignoni G, Netto G, Argani P. Perivascular epithelioid cell tumors (PEComas) harboring TFE3 gene rearrangements lack the TSC2 alterations characteristic of conventional PEComas: further evidence for a biological distinction. Am J Surg Pathol. 2012;36(5):783–4.
65. Pan CC, Chung MY, Ng KF, Liu CY, Wang JS, Chai CY, et al. Constant allelic alteration on chromosome 16p (TSC2 gene) in perivascular epithelioid cell tumour (PEComa): genetic evidence for the relationship of PEComa with angiomyolipoma. J Pathol. 2008;214(3):387–93.
66. Sanfilippo R, Jones RL, Blay JY, Le Cesne A, Provenzano S, Antoniou G, et al. Role of chemotherapy, VEGFR inhibitors, and mTOR inhibitors in advanced perivascular epithelioid cell tumors (PEComas). Clin Cancer Res. 2019;25(17):5295–300.
67. Soslow RA, Ali A, Oliva E. Mullerian adenosarcomas: an immunophenotypic analysis of 35 cases. Am J Surg Pathol. 2008;32(7):1013–21.

68. Gallardo A, Prat J. Mullerian adenosarcoma: a clinicopathologic and immunohistochemical study of 55 cases challenging the existence of adenofibroma. Am J Surg Pathol. 2009;33(2):278–88.
69. Nathenson MJ, Ravi V, Fleming N, Wang WL, Conley A. Uterine Adenosarcoma: a Review. Curr Oncol Rep. 2016;18(11):68.
70. Howitt BE, Sholl LM, Dal Cin P, Jia Y, Yuan L, MacConaill L, et al. Targeted genomic analysis of Müllerian adenosarcoma. J Pathol. 2015;235(1):37–49.
71. Blom R, Guerrieri C. Adenosarcoma of the uterus: a clinicopathologic, DNA flow cytometric, p53 and mdm-2 analysis of 11 cases. Int J Gynecol Cancer. 1999;9(1):37–43.
72. Hodgson A, Howitt BE, Park KJ, Lindeman N, Nucci MR, Parra-Herran C. Genomic characterization of HPV-related and gastric-type Endocervical adenocarcinoma: correlation with subtype and clinical behavior. Int J Gynecol Pathol. 2020;39(6):578–86.
73. Mills AM, Karamchandani JR, Vogel H, Longacre TA. Endocervical fibroblastic malignant peripheral nerve sheath tumor (neurofibrosarcoma): report of a novel entity possibly related to endocervical CD34 fibrocytes. Am J Surg Pathol. 2011;35(3):404–12.
74. Chiang S, Cotzia P, Hyman DM, Drilon A, Tap WD, Zhang L, et al. NTRK fusions define a novel uterine sarcoma subtype with features of Fibrosarcoma. Am J Surg Pathol. 2018;42(6):791–8.
75. Croce S, Hostein I, Longacre TA, Mills AM, Pérot G, Devouassoux-Shisheboran M, et al. Uterine and vaginal sarcomas resembling fibrosarcoma: a clinicopathological and molecular analysis of 13 cases showing common NTRK-rearrangements and the description of a COL1A1-PDGFB fusion novel to uterine neoplasms. Mod Pathol. 2019;32(7):1008–22.
76. Kirsch CH, Goodman M, Esiashvili N. Outcome of female pediatric patients diagnosed with genital tract rhabdomyosarcoma based on analysis of cases registered in SEER database between 1973 and 2006. Am J Clin Oncol. 2014;37(1):47–50.
77. Nasioudis D, Alevizakos M, Chapman-Davis E, Witkin SS, Holcomb K. Rhabdomyosarcoma of the lower female genital tract: an analysis of 144 cases. Arch Gynecol Obstet. 2017;296(2):327–34.
78. Daya DA, Scully RE. Sarcoma botryoides of the uterine cervix in young women: a clinicopathological study of 13 cases. Gynecol Oncol. 1988;29(3):290–304.
79. Ferguson SE, Gerald W, Barakat RR, Chi DS, Soslow RA. Clinicopathologic features of rhabdomyosarcoma of gynecologic origin in adults. Am J Surg Pathol. 2007;31(3):382–9.
80. Pinto A, Kahn RM, Rosenberg AE, Slomovitz B, Quick CM, Whisman MK, et al. Uterine rhabdomyosarcoma in adults. Hum Pathol. 2018;74:122–8.
81. de Kock L, Boshari T, Martinelli F, Wojcik E, Niedziela M, Foulkes WD. Adult-onset cervical embryonal rhabdomyosarcoma and DICER1 mutations. J Low Genit Tract Dis. 2016;20(1):e8–e10.
82. Doros L, Yang J, Dehner L, Rossi CT, Skiver K, Jarzembowski JA, et al. DICER1 mutations in embryonal rhabdomyosarcomas from children with and without familial PPB-tumor predisposition syndrome. Pediatr Blood Cancer. 2012;59(3):558–60.
83. Sorensen PH, Lynch JC, Qualman SJ, Tirabosco R, Lim JF, Maurer HM, et al. PAX3-FKHR and PAX7-FKHR gene fusions are prognostic indicators in alveolar rhabdomyosarcoma: a report from the children's oncology group. J Clin Oncol. 2002;20(11):2672–9.
84. Parham DM, Barr FG. Classification of rhabdomyosarcoma and its molecular basis. Adv Anat Pathol. 2013;20(6):387–97.

# Chapter 5
# Clinical Relevance of *BRCA1/2* Pathogenic Variants and Impaired DNA Repair Pathways in Ovarian Carcinomas

**Akira Nishijima, Michihiro Tanikawa, and Katsutoshi Oda**

**Abstract** Genetic analysis of *BRCA1/2* pathogenic variants has become indispensable in clinics, especially for advanced ovarian cancer patients. The frequency of *BRCA1/2* pathogenic variants is significantly higher in high-grade serous ovarian carcinomas (HGSOC), and the clinical benefit of PARP inhibitors was proven in many clinical trials, whose patients were predominantly HGSOC. As tumor-only NGS analysis, including Myriad myChoice® CDx, has been commonly performed, distinction of germline and somatic pathogenic variants is needed in patients whose tumors harbor *BRCA1/2* pathogenic variants. As sensitivity of both PARP inhibitors and platinum chemotherapy is deeply associated with homologous recombination deficiency, the knowledge of DNA repair pathways has become essential in the treatment of ovarian cancer. We discuss mutational analysis of *BRCA1/2* and other homologous recombination pathway genes, DNA repair pathways for PARP inhibitors and platinum chemotherapy, genomic instability analysis for detection of homologous recombination deficiency, and reversion mutations of *BRCA1/2* from cell-free DNA in liquid biopsy.

**Keywords** BRCA1/2 · DNA repair · Homologous recombination deficiency · Poly (ADP-ribose) polymerase inhibitor · Companion diagnostics · Reversion mutation

A. Nishijima · M. Tanikawa
Department of Obstetrics and Gynecology, Graduate School of Medicine, The University of Tokyo, Bunkyo City, Tokyo, Japan

K. Oda (✉)
Division of Integrative Genomics, Graduate School of Medicine, The University of Tokyo, Bunkyo City, Tokyo, Japan
e-mail: katsutoshi-tky@umin.ac.jp



M. Mandai (ed.), *Personalization in Gynecologic Oncology*, Comprehensive Gynecology and Obstetrics, https://doi.org/10.1007/978-981-19-4711-7_5

## 5.1 Introduction

High-grade serous ovarian carcinoma (HGSOC) is the most common histology type in ovarian carcinomas. The ratio of stage III/V is high, and the prognosis is poor in HGSOC. An important risk factor for HGSOC is germline *BRCA1* or *BRCA2* pathogenic variants (i.e., gBRCApv), known as hereditary breast and ovarian cancer (HBOC) [1–3]. As well, somatic *BRCA1/2* pathogenic variants (i.e. sBRCApv) are commonly observed in various types of cancers, including HGSOC [4–6]. Therefore, the distinction of germline and somatic pathogenic variants is needed to appropriately diagnose HBOC when *BRCA1/2* pathogenic variants are detected in tumor tissues (tBRCApv).

Both BRCA1 and BRCA2 are indispensable in DNA double-strand break (DSB) repair by homologous recombination (HR) [7–9]. The poly(ADP-ribose) polymerase (PARP) is an essential enzyme for the base excision repair (BER) pathway and PARP inhibitors cause synthetic lethality in cancer cells with HR deficiency (HRD) [10, 11]. Indeed, gBRCApv is shown to be a strong biomarker of various PARP inhibitors in ovarian carcinoma, as well as other HBOC-related (breast, prostate, and pancreatic) cancers [12–14]. In this review, we focus on *BRCA1/2* pathogenic variants, HRD status, DNA repair pathways, and their clinical relevance in ovarian cancer.

## 5.2 Mutational Analysis of *BRCA1/2* (gBRCApv, sBRCApv, and tBRCApv)

The ratio of gBRCApv is detected at 10–15% of ovarian cancer, and at 14–28% in HGSOC [3, 4]. gBRCA genetic testing with blood samples (such as BRACAnalysis CDx) is directly linked to the diagnosis of HBOC and is also used as companion diagnostics (CDx) for maintenance treatment with olaparib. Since 2020, gBRCA genetic testing was covered by health insurance in any ovarian cancer patients in Japan, regardless of timing and histologic types. Therefore, the status of gBRCApv is more broadly evaluated, and can be used as a biomarker for PARP inhibitors (either olaparib or niraparib) in platinum-sensitive, relapsed ovarian cancer. The ratio of gBRCApv in endometrioid, clear cell, mucinous and low-grade serous carcinomas was reported to be 6.7% (8/120), 2.1% (4/187), 0% (0/19), and 20% (1/5) in Japan Charlotte study [3]. A meta-analysis also supported the low ratio of gBRCApv in clear cell (3%) and mucinous (2.5%) ovarian carcinomas [15].

tBRCApv can be analyzed by using tumor tissues (or serum by liquid biopsy), however, the distinction between gBRCApv and sBRCApv is not confirmatory

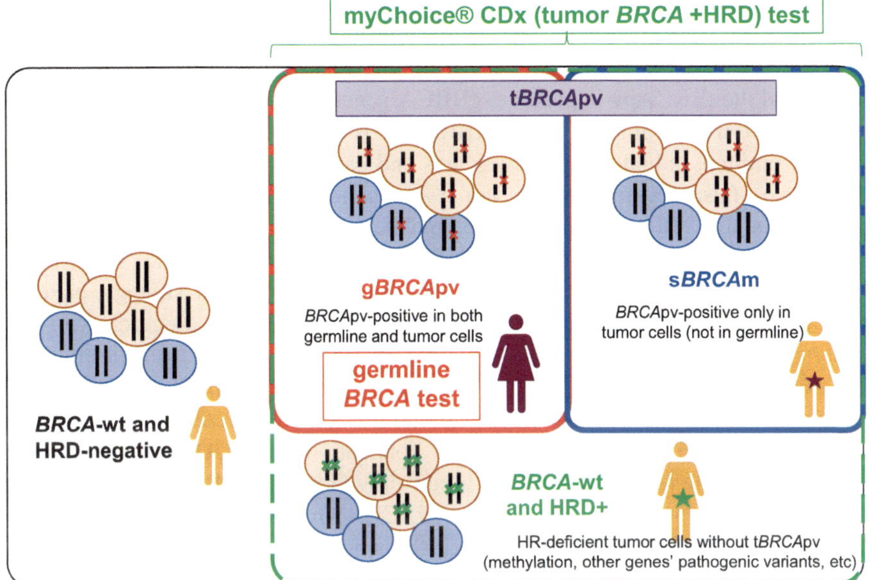


**Fig. 5.1** *Classification by BRCApv and HRD in ovarian carcinomas.* Germline *BRCA* test only verifies gBRCApv, while MyChoice®® CDx test analyzes both tBRCApv and HRD. gBRCApv and sBRCApv are usually mutually exclusive, and either gBRCApv or sBRCApv is accompanied by loss of heterozygosity (LOH) of the opposite allele. HRD-positive tumors without tBRCApv should have other mechanisms of HRD, instead of tBRCApv

(Fig. 5.1). Therefore, gBRCApv assessment is needed to definitely diagnose sBRCApv. myChoice® CDx from Myriad is a test to evaluate tBRCApv, as well as HRD status by genomic instability scores [16, 17], and is approved as CDx for primary, advanced ovarian carcinomas (and platinum-sensitive ovarian carcinomas with 3 or more regimens) in Japan (Fig. 5.1). Information about the exact ratio of sBRCApv is limited, compared with that of gBRCApv or tBRCApv, due to the necessity of assessment with both tumor and normal DNA. The ratio of sBRCApv is estimated to be 5–7%, which is compatible with the first whole-exome report from the Cancer Genome Atlas (TCGA) Research Network [4, 18, 19]. ASCO guideline recommended that all epithelial ovarian cancer patients should be offered gBRCA genetic testing and other ovarian cancer susceptibility genes, irrespective of their clinical features or family cancer history [20]. Moreover, testing for tBRCApv should be recommended in ovarian cancer patients, even without carrying a gBRCApv [20]. The concordance rate of *BRCA* pathogenic variant status in tumor (tBRCApv) versus blood (gBRCApv) was high in phase III SOLO2 trial [21]. Among 241 patients, there was 98% and 100% concordance between tumor and germline testing for BRCA1pv and BRCA2pv, respectively, and discordance was found only in four patients (2%) [21]. Considering the high concordance rate, tBRCApv testing may be performed prior to gBRCApv testing. When MyChoice® CDx is initially tested and the tBRCApv is positive, gBRCA genetic testing is needed to diagnose gBRCApv (HBOC) or sBRCApv. If tBRCApv is negative, gBRCA genetic testing may be omitted, although both physicians and patients should realize that discordance (i.e., false negative) rate is not 0%. When gBRCA genetic testing is performed first and gBRCApv is negative, myChoice® CDx testing would be needed to evaluate the status of sBRCApv and HRD. As the main aim of CDx for primary advanced ovarian cancer patients is to determine the treatment strategy, patients may prefer to take myChoice® CDx testing in the beginning. Regardless of the order and the timing of each test, appropriate informed consent and genetic counseling should be ensured for all the patients who are potentially HBOC.

Another approach to assess the mutational status of *BRCA1/2* is comprehensive genomic profiling (CGP) tests, such as FoundationOne® CDx Comprehensive genomic profiling [22–24]. Although tumor-only (T-only) CGP is more frequently tested, a tumor-normal paired (T-N paired) CGP, OncoGuide™ NCC Oncopanel system, is also available in Japan, which can simultaneously diagnose both gBRCApv and sBRCApv [25–28]. The other T-N paired CGP, named Todai OncoPanel, which is composed of both DNA and RNA panel, has also been in progress for approval from Pharmaceuticals and Medical Devices Agency (PMDA) in Japan [29, 30]. TruSight Oncology 500 (TSO500) Assay is another T-only CGP panel [31], which is currently examined in solid tumors under Advanced Medical Care Category B for future PMDA approval in Japan. Liquid biopsy is the other option as CGP testing, including FoundationOne Liquid® CDx and Guardant360 [32, 33]. Liquid biopsy focuses on cell-free circulating tumor DNA (ctDNA) to identify pathogenic variants in tumors, but can also identify germline variants. Therefore, liquid biopsy alone cannot be a direct test of gBRCApv, as same as T-only panel. However, the variant allele frequency (VAF) from ctDNA is often much lower than the VAF from germline variants, which may make us more easily assume presumed gBRCApv. Indeed, presumed gBRCApv has been identified in various types of cancers from ctDNA analysis in liquid biopsy [34]. From the point of view of ovarian cancer treatment, the utility of CGP in Japan is still limited for both diagnosis of HBOC and CDx for PARP inhibitors, as the CGP testing under health insurance coverage can be applied only to cancer patients who have (already or almost) finished all the standardized treatments. This means that most ovarian cancer patients can receive CGP tests only after acquisition of platinum resistance. Therefore, gBRCApv testing and/or myChoice® CDx are preferentially performed beforehand of the CGP tests, if available. Biomarkers and CDx for PARP inhibitors in ovarian carcinomas are summarized in Table 5.1, including information about treatment regimen and clinical applications (such as SOLO-1, SOLO-2, PAOLA-1, Study 19/OPINION, PRIMA, NOVA, QUADRA, and OReO) [35–43].

**Table 5.1** Biomarkers and CDx for PARP inhibitors in ovarian carcinomas

| Biomarker (Laboratory)/CDx | Specimen | Types of test | Ratio (HGSOC) | CDx | Term to be tested (in Japan) | Regimen |
|---|---|---|---|---|---|---|
| g*BRCA*m | Blood | Germline *BRCA1/2* genetic test (BRACAnalysis) | 14–28% | ○ | 1st line for CDx, anytime for HBOC diagnosis | SOLO-1, PAOLA-1 |
| t*BRCA*m by CGP (g*BRCA*m + s*BRCA*m) | Tumor (FFPE) | FoundationOne CDx etc | 30% (s*BRCA*m included) | ○ | After standardized treatment for CGP | N/A (as examined after standardized treatment) |
| LOH score (biomarker to evaluate HRD, i.e. genomic instability) | Tumor (FFPE) | FoundationOne CDx | 63% (including t*BRCA*m at 35%) | N/A | After standardized treatment for CGP | |
| HRD (tBRCAm+GIS: Genomic instability score) | Tumor (FFPE) | Myriad myChoice CDx | 50% (including t*BRCA*m at 30%) | ○ | 1st line as maintenance therapy, 3 or more lines as monotherapy | PAOLA-1 (first line), QUADRA (3 or more lines) |
| Biomarker (clinical) | Specimen | Types of test | Ratio (HGSOC) | CDx | Term to be tested (in Japan) | Regimen (not as CDx) |
| Platinum-sensitivity (CR/PR) | (–) | RECIST etc | 70–80% (first line) | N/A | During treatment by platinum-based chemotherapy | PRIMA, SOLO-1,, PAOLA-1, NOVA, study 19 (OPINION), SOLO-2, QUADRA, OReO |

Regimen and applications: (1) SOLO-1, olaparib maintenance therapy, first line for gBRCApv/tBRCApv, (2) PAOLA-1, olaparib and bevacizumab maintenance therapy, first line for tBRCApv and/or HRD-positive. (3) QUADRA, niraparib monotherapy, recurrent (after 3 or more regimen), platinum-sensitive and HRD-positive, (4) PRIMA, niraparib maintenance therapy, first line and platinum-sensitive. (5) NOVA, niraparib maintenance therapy, recurrent and platinum-sensitive, (6) Study 19 (OPINION), olaparib maintenance therapy, recurrent and platinum-sensitive. (7) SOLO-2, olaparib maintenance therapy, recurrent and platinum-sensitive, (8) OReO, olaparib maintenance therapy (rechallenge), recurrent and platinum-sensitive (niraparib use also allowed)

## 5.3 DNA Repair Pathway for Single-Strand and Double-Strand Breaks

Cells have diverse DNA damage response pathways to protect the genome and maintain genomic homeostasis. Single-strand breaks (SSBs) are easily caused by endogenous DNA damage (i.e., methylation, cytosine deamination, and oxygen radicals), UV light and any other genotoxic agents [44]. SSBs can be repaired by base excision repair (BER). PARP-1 binds to SSBs and activates synthesis of poly(ADP-ribose) polymers, which subsequently allows BER proteins to access the SSB and complete the BER process [45]. DSBs can be generated by either endogenous sources (such as reactive oxygen species) or exogenous sources (such as ionizing radiation and chemotherapeutic agents). In addition, DSBs are also generated by replication-associated errors, during meiosis, and unrepaired SSBs [46]. Double-strand breaks (DSBs) are highly lethal to cells, and the types of DSB repair pathways are crucial to cells to survive (Table 5.2) [46, 47]. DSB repair pathways

**Table 5.2** DNA repair pathways

| DNA damage | Single-strand break (SSB) | Double-strand break (DSB) | | | | Interstrand crosslink (ICL) |
|---|---|---|---|---|---|---|
| Causal factors | Endogenous, UV-light, Genotoxic agent, etc | Meiosis, ROS, X-rays, Unrepaird SSB etc. | | | | Genotoxic agent (cisplatin, etc) |
| Types of repair | SSB repair | DSB repair | | | | ICL pathway |
| DNA repair pathways | Base excision repair | NHEJ | Alt-NHEJ | SSA | HR | FA pathway<br>Nucleotide excision repair<br>Translesion synthesis<br>HR |
| Key molecules | PARP-1 | Ku80-Ku70 heterodimer | PARP-1 | CtIP | ATM, ATR | FANCM–FAAP24–MHF1/2 complex |
| | OGG1 | DNA-PK | MRN/CtIP complex | MRN complex | MRN complex | FANCA/E/D2/I |
| | XRCC1 | Atemis | XRCC1 | XPF/ERCC | BRCA1/2, FANCD | FANCP/SLX4 |
| | Polβ | XRCC4 | RAD52 | RAD52 | RAD51, FANCF | REV1, pol ζ. |
| | PCNA | Polε | Polθ | | Polδ, Polε | BRCA1/2, RAD51 |
| | DNA ligase III | DNA ligase IV | DNA ligase III | | DNA ligase I | And other HR pathway molecules |

include canonical nonhomologous end-joining (C-NHEJ), alternative NHEJ (Alt-NHEJ), single-strand annealing (SSA), and homologous recombination (HR). In addition to c-NHEJ pathway, Alt-NHEJ pathway has been proven to be a leading actor [48, 49]. C-NHEJ is an error-prone repair pathway, which does not require a DNA template and is mediated by the direct joining of the two broken ends [46]. Alt-NHEJ repair, also known as microhomology-mediated end joining (MMEJ) pathway, requires from 2 to 20 nucleotides of sequence homology at DNA ends of DSBs to start repair [48, 49]. SSA requires annealing of a relatively long tract (more than 20–50 bp) of complementary sequence in resected overhangs [49–53]. As both Alt-NHEJ and SSA do not operate with a DNA template strand (sister chromatid), they cannot restore original DNA sequence, and are error prone [48, 50]. These error-prone repair of DSBs can result in miss joining of DNA repair components with DNA ends, which leads to genomic instability with various deletions, inversions, or complex rearrangements of chromosomes (i.e., genomic scar) [46, 51–53]. In contrast, HR requires a homologous DNA template (sister chromatid) to start the repair. Only HR has the potential to be restorative, as the template for recombination is the identical sister chromatid [51–53]. In summary, HRD is highly associated with genomic instability, due to the dependency on non-HR DSB repair pathways.

## 5.4 Cytotoxicity of PARP Inhibitors and Platinum Chemotherapy

PARP-1 is indispensable for both BER and Alt-NHEJ. Inhibition of PARP causes DSB accumulation by blocking BER of SSBs [52]. In HRD tumor cells, DSB should be repaired without HR, which leads to dependency on error-prone repair pathways, such as NHEJ and SSA (Table 5.2). Therefore, PARP inhibitors are theoretically effective to HRD cancer cells.

DSBs are also induced during the repair process of DNA interstrand cross-link (ICL), which is generated by platinum chemotherapy [46, 54]. To repair ICL, various types of DNA repair pathways are indispensable, including Fanconi Anemia (FA) pathway, nucleotide excision repair (NER) pathway, Translesion synthesis (TLS) pathway, as well as HR pathway (Table 5.2) [54–57]. The Fanconi anemia (FA) family includes 19 distinct functional complementation groups (A, B, C, D1, D2, E, F, G, I, J, L, M, N, O, P, Q, R, S, T), and *BRCA1*, *BRCA2* and other HR repair genes, such as *BRIP1* and *RAD51*, encode a subset of FA proteins. FA gene products play key roles in both inhibiting ICL formation and repairing the ICL [51, 54] (Table 5.3). FA pathway in ICL repair includes the processes of recognition of ICL by BRCA1 and RAD51, recruitment of the core FA complex, and ubiquitination of FA core complex (FANCI/D2). Then, endonucleolytic incision is mediated by SLX4 (FANCP), which recruits multiple nucleases (ERCC1-XPF, SLX1, and MUS81-EME) in the NER pathway. Nucleolytic incisions unhook the ICL, and the defective lesion can be recovered by translesion synthesis, mediated by REV1 or Poltheta polymerases, to generate templates for HR repair. Finally, HR repair is conducted by using the templates [54, 58, 59]. As both BRCA1 and

**Table 5.3** Fanconi Anemia genes and main functions

| Gene | Synonym | Main protein functions |
|---|---|---|
| *FANCA* | | Component of FA core complex |
| *FANCB* | | |
| *FANCC* | | |
| *FANCD1* | *BRCA2* | HR repair; loads RAD51 onto DNA; interacts with FANCD2 and FANCN; stalled replication fork protection |
| *FANCD2* | | Ubiquitinated after DNA damage; MCM interaction; stalled replication fork protection |
| *FANCE* | | Component of FA core complex |
| *FANCF* | | |
| *FANCG* | *XRCC9* | |
| *FANCI* | | Ubiquitinated after DNA damage; activates dormant origins |
| *FANCJ* | *BACH, BRIP1* | FA repair; HR repair; 3′ to 5′ helicase; interacts with BRCA1; checkpoint activation |
| *FANCL* | | Component of FA core complex |
| *FANCM* | | DNA helicase/translocase; localises the core complex to DNA |
| *FANCN* | *PALB2* | HR repair; promotes BRCA2 function; interacts with BRCA1 and BRCA2 |
| *FANCO (provisional)* | *RAD51C* | HR repair; promotes RAD51 nucleoprotein filament stability; ICL repair |
| *FANCP* | *SLX4* | Coordinates XPF–ERCC1, MUS81–EME1 and SLX1 nucleases; resolves Holliday junctions |
| *FANCQ* | *ERCC4, XPF* | Endonuclease; binds to ERCC1; crosslink unhooking |
| *FANCR* | *RAD51* | HR repair; ICL repair; protection of nascent strands from DNA2- and WRN-mediated resection; stalled replication fork protection |
| *FANCS* | *BRCA1* | HR repair; promotes RAD51 loading; ICL repair; chromatin dissociation of replicative helicase; stalled replication fork protection; interacts with FANCD2 and FANCN |
| *FANCT* | *UBE2T* | E2 ubiquitin-conjugating enzyme for FANCI–FANCD2 complex; interacts with FANCL |

BRCA2 cooperate with other factors to repair ICL, as well as HR, ICL deficiency by tBRCApv may be also associated with platinum sensitivity.

The effect of PARP inhibitors is not limited to their catalytic activities (i.e. accumulation of DSBs through unrepaired SSBs). PARP–DNA complexes by PARP trapping interfere with DNA replication, transcription, and repair, resulting in replication fork collapse and cell death [53, 60–62]. PARP trapping is more toxic than the accumulation of unrepaired SSBs, and the potency of PARP trapping is distinct among the PARP inhibitors [60, 61]. The trapping potencies of PARP inhibitors are the strongest in Talazoparib, and the weakest in Veliparib [60]. The potencies are comparable among Niraparib, Olaparib, and Rucaparib, but the potency of Niraparib was reported to be stronger than that of Olaparib and Rucaparib. As the mean half-life is also longer in Niraparib, the PARP trapping potency may be more robustly maintained in Niraparib compared with Olaparib and Rucaparib. As PARP–DNA complex is more toxic and cannot be repaired by HR alone, PARP inhibitors may be effective for HR-proficient cancer cells. Indeed, Niraparib maintenance therapy was shown to improve progression-free survival of HR-proficient ovarian cancers in both the first line setting and the recurrent setting with platinum sensitivity [36, 41]. Taken together, repair of PARP–DNA complex may need some common processes with repair of ICL. Further study is warranted to fully understand the molecular mechanism of PARP inhibition.

## 5.5 PARP Inhibitors in *BRCA1/2* Mutated Ovarian Carcinomas

Before being established as an HRD biomarker, gBRCApv was broadly tested in HGSOC, especially with a family history of breast and/or ovarian cancer to diagnose HBOC [63, 64]. Olaparib monotherapy was first approved by FDA (U.S. Food and Drug Administration) in 2014 in patients with deleterious gBMCAm ovarian cancer (with three or more lines of chemotherapy) [65, 66]. Two phase 3 trials of olaparib maintenance therapy (SOLO1 and SOLO2) enrolled exclusively gBRCApv patients and showed significant improvement of PFS in gBRCApv patients with high-grade serous or endometrioid ovarian cancer in complete or partial response to platinum-based chemotherapy (HR 0.30, 0.23–0.41, and HR 0.30, 0.22–0.41, respectively) [35, 37]. The prognostic impact of gBRCApv has also been shown in clinical trials with other PARP inhibitors (niraparib, rucaparib, and veliparib) [36, 41, 67, 68]. Thus, gBRCApv has been the main target of PARP inhibitors. Up to now, Olaparib has been approved only in patients with breast, prostate and pancreatic carcinomas with BRCApv (either gBRCApv and/or tBRCApv) in Japan [4, 12]. In ovarian carcinomas, all the regimens in the clinical trials with Olaparib (SOLO-1, PAOLA-1, Study 19 (OPINION), and SOLO-2) or Niraparib (PRIMA, NOVA, and QUADRA) have been approved in Japan, whereas regimen with Rucaparib and Veliparib are not [35–43]. Although the number of sBRCApv in clinical trials with a PARP inhibitor was small ($n$ = 20/209, 9.6% in Study 19 and $n$ = 47/553, 8.5% in NOVA trial), sBRCApv group showed favorable prognosis, compared with the placebo group (HR 0.23, 0.04–1.12 in Study 19 and HR 0.27, 0.08–0.90 in NOVA trial) [35, 36, 69]. Therefore, tBRCApv, as well as gBRCApv, is currently tested as a CDx for PARP inhibitors in ovarian cancer.

## 5.6 Genomic Scar (Chromosomal Rearrangements) Assays for Genomic Instability

Chromosomal aberrations are induced by defective DNA repair, and gross chromosomal rearrangements are associated with HRD [69, 70]. Such genomic scar, represented by LOH, can be scored by using SNP array data (or array CGH) [71]. The Genomic Instability score by MyChoice® CDx includes loss of heterozygosity

(LOH), telomeric allele imbalance (telomeric allelic The sum of the three scores is 42 or more to be considered HRD positive [16].

FoundationOne CDx LOH Score was used in the ARIEL-2 and ARIEL-3 studies of the PARP inhibitor rucaparib in platinum-sensitive patients with recurrent ovarian cancer. Initially, the threshold was set at 14% or higher as HRD positivity (ARIEL-2), but in ARIEL-3, the LOH score threshold was set at 16% or higher for analysis [67, 72]. Although the case backgrounds differ, and further validation of the consistency between the two tests is desirable, both tests are useful for estimating HRD. Currently, the LOH score in the FoundationOne CDx is not covered by insurance as a cCDx, and is only listed as a reference value in the report of the CGP test for ovarian carcinomas. If the timing of cancer gene panel testing could be accelerated, the data would be more helpful in considering the introduction of PARP inhibitors (even if they cannot be used as CDx).

## 5.7 Mutational Analysis of HRR-Related Genes

HR repair needs multiple steps and multiple molecules, pathogenic variants in other HR repair (HRR) genes are also important. In addition, epigenetic changes (hypermethylation in the promoter region) are also reported to be associated with HRD in ovarian cancer [4]. The FDA approval of olaparib in prostate cancer covers the following HRR pathogenic variants: *BRCA1/2*, *ATM*, *BARD1*, *BRIP1*, *CDK12*, *CHEK1*, *CHEK2*, *FANCL*, *PALB2*, *RAD51B*, *RAD51C*, *RAD51D* and *RAD54L* [14], although the approval is limited to those with gBRCApv and/or tBRCApv in Japan. Ratio of deleterious germline pathogenic variants in *BRIP1* (1.36%), *PALB2* (0.62%), *RAD51C* (0.57%), *RAD51D* (0.57%), and *BARD1* (0.21%) were all significantly higher in ovarian cancer patients (n = 1345, including 1118 HGSOC) than in control [73]. Germline pathogenic variants in *CHEK2*, *ATM*, and *NRN* were also identified at 0.47–0.57% in this study [73]. In triple-negative breast cancers (*n* = 1824), 3.7% of deleterious germline pathogenic variants were identified in 15 non-BRCA predisposition genes, including *PALB2* (1.2%), *BARD1*, *RAD51D*, *RAD51C*, and *BRIP1* [74]. As the ratio of other HRRm was low (germline pathogenic variants at 6.0% and somatic pathogenic variants at 2.7%) in 390 ovarian carcinomas [4], other HRRm status has not been established as a biomarker to PARP inhibitors. However, other HRRm was associated with prognosis in several retrospective studies. Other HRRm group showed longer progression-free survival (PFS) by olaparib maintenance therapy, compared with the group with no detectable BRCApv or other HRRm in Study 19 (hazard ratio:HR 0.21, 0.04–0.86) [75]. Adjusted hazard for PFS was lower in patients with other HRRm (*n* = 81 of 1195: 6.8%) in 14 genes (*ATM*, *ATR*, *BARD1*, *BLM*, *BRIP1*, *CHEK2*, *MRE11A*, *NBN*, *PALB2*, *RAD51C*, *RAD51D*, *RBBP8*, *SLX4* and *XRCC2*), compared with patients with no HRRm (HR 0.73, 0.57–0.94) in GOG-218 study (a randomized phase III trial in advanced ovarian carcinoma of bevacizumab added to paclitaxel and carboplatin) [76].

Loss of function of *BRCA1* and *RAD51C* by hyper-methylation is also frequently observed in ovarian cancer [4, 77, 78]. The ratio of methylation of BRCA1 and RAD51C is 6–11% and 3% in HGSOC, respectively [4, 78]. Epigenetic silencing of these genes was mutually exclusive with BRCApv and other HRRm, and was considered as one of the mechanisms of HRD in HGSOC [4]. However, prognosis of patients with epigenetic silencing of *BRCA1* was not as favorable as those with BRCApv [4, 79]. Therefore, hypermethylation may not be functionally equivalent to BRCApv in mediating platinum sensitivity. One possible reason may be reversion from hypermethylation to hypomethylation in relapsed tumors. Hypermethylation status of *BRCA1* (or *RAD51C*) may be often modified by the platinum-based chemotherapy.

## 5.8 Reversion Mutation

Acquired (secondary) *BRCA* reversion mutations are key resistance mechanism to platinum-based chemotherapies and PARP inhibitors [80]. Reversion mutations in *BRCA* should recover the function of BRCA, which is distinct from other resistant pathogenic variants in oncogenes, such as *EGFR*. The method to detect *BRCA* reversion mutations is to analyze cfDNA by liquid biopsy, including FoundationOne Liquid CDx and Guardant 360 [81]. Taking advantage of liquid biopsy in sample availability and detection of heterogeneity, reversion mutations of *BRCA* have been reported in ovarian cancer patients who were treated with PARP inhibitors [82, 83]. Notably, multiple reversion mutations can happen in the same patient, suggesting the heterogeneity of resistant clones [80, 83]. Currently, lower limit of detection of VAF is approximately 0.4% or lower for substitutions and insertions/deletions in FoundationOne Liquid CDx, which may fail to detect reversion mutations from minor clones [32]. cfDNA analysis was reported in ovarian cancer patients with deleterious germline or somatic *BRCA* pathogenic variants treated with a PARP inhibitor, rucaparib [80]. The ratio of *BRCA* reversion mutations identified in tBRCApv ovarian cancer patients in ARIEL2 study was 18% (2/11) from pretreatment, 13% (5/38) from platinum-refractory, and 2% (1/48) from platinum-sensitive cfDNA [80]. Progression free survival by Rucaparib was significantly shorter in the group with reversion mutations (median 1.8 months), suggesting the recovery of HR repair pathway [80]. According to a meta-analysis, the ratio of reversion mutations in patients with tBRCApv on progression after platinum or PARPi treatment across the different tumor types was 26.0%, and the ratio in ovarian cancers was 21.4% (33/154) in *BRCA1* mutated and 27.8% (22/80) in *BRCA2* mutated patients [83]. A Phase IIIb trial, OReO/ENGOT Ov-38, with maintenance olaparib rechallenge in ovarian cancer patients previously treated with a PARP inhibitor, indicated that PFS was significantly improved by olaparib rechallenge (Median PFS was 4.3 months in olaparib arm vs 2.8 m in placebo arm: hazard ratio 0.57; 95% CI 0.37–0.87; P = 0.022) [42]. However, reversion mutations were not evaluated, and it is less likely that olaparib rechallenge was effective in the patients with *BRCA* reversion mutations. Real world data of the reversion mutations by liquid biopsy analysis is highly warranted to appropriately address the clinical impact of PARP rechallenge.

## 5.9 Conclusion

Although the functions of *BRCA1/2* genes and other HRR genes have been well studied, clinical applications of PARP inhibitors (including the regimen with other molecular targeted drugs) convey various clinical questions to be solved. The questions include pharmacodynamics and effect of PARP trapping by each PARP inhibitor, mechanism to repair PARP-DNA complex, efficacy of PARP inhibitors in HR-proficient tumors, mechanism of reversion mutations of *BRCA*, and mechanism of acquired resistance in non-BRCApv tumors. Combination therapy of PARP inhibitors and other drugs (such as antiangiogenic inhibitors and immune checkpoint inhibitors) also highlights the significance of the molecular mechanism of PARP inhibitors in ovarian carcinomas.

**Acknowledgements** We would like to thank Editage (www.editage.com) for English language editing.

**COI Disclosure** Dr. Oda received a research grant from Konica Minolta, INC. and AstraZenea plc., and lecture fees from Chugai Pharmaceutical Co., Ltd., AstraZeneca plc, Takeda Pharmaceutical Co., Ltd.

## References

1. Pal T, Permuth-Wey J, Betts JA, Krischer JP, Fiorica J, Arango H, LaPolla J, Hoffman M, Martino MA, Wakeley K, Wilbanks G, Nicosia S, Cantor A, Sutphen R. BRCA1 and BRCA2 mutations account for a large proportion of ovarian carcinoma cases. Cancer. 2005;104:2807–16.
2. Daly MB, Pilarski R, Yurgelun MB, Berry MP, Buys SS, Dickson P, Domchek SM, Elkhanany A, Friedman S, Garber JE, Goggins M, Hutton ML, Khan S, Klein C, Kohlmann W, Kurian AW, Laronga C, Litton JK, Mak JS, Menendez CS, Merajver SD, Norquist BS, Offit K, Pal T, Pederson HJ, Reiser G, Shannon KM, Visvanathan K, Weitzel JN, Wick MJ, Wisinski KB, Dwyer MA, Darlow SD. NCCN Guidelines Insights: Genetic/Familial High-Risk Assessment: Breast, Ovarian, and Pancreatic, Version 1.2020. J Natl Compr Canc Netw. 2020;18:380–91.
3. Enomoto T, Aoki D, Hattori K, Jinushi M, Kigawa J, Takeshima N, Tsuda H, Watanabe Y, Yoshihara K, Sugiyama T. The first Japanese nationwide multicenter study of BRCA mutation testing in ovarian cancer: CHARacterizing the cross-sectionaL approach to ovarian cancer geneTic TEsting of BRCA (CHARLOTTE). Int J Gynecol Cancer. 2019;29:1043–9.
4. The Cancer Genome Atlas Research Network. Integrated genomic analyses of ovarian carcinoma. Nature. 2011;474:609–15.
5. Pennington KP, Walsh T, Harrell MI, Lee MK, Pennil CC, Rendi MH, Thornton A, Norquist BM, Casadei S, Nord AS, Agnew KJ, Pritchard CC, Scroggins S, Garcia RL, King MC, Swisher EM. Germline and somatic mutations in homologous recombination genes predict platinum response and survival in ovarian, fallopian tube, and peritoneal carcinomas. Clin Cancer Res. 2014;20:764–75.

6. Moschetta M, George A, Kaye SB, Banerjee S. BRCA somatic mutations and epigenetic BRCA modifications in serous ovarian cancer. Ann Oncol. 2016;27:1449–55.
7. Farmer H, McCabe N, Lord CJ, Tutt AN, Johnson DA, Richardson TB, Santarosa M, Dillon KJ, Hickson I, Knights C, Martin NM, Jackson SP, Smith GC, Ashworth A. Targeting the DNA repair defect in BRCA mutant cells as a therapeutic strategy. Nature. 2005;434:917–21.
8. Banerjee S, Kaye SB, Ashworth A. Making the best of PARP inhibitors in ovarian cancer. Nat Rev Clin Oncol. 2010;7:508–19.
9. Oda K, Tanikawa M, Sone K, Mori-Uchino M, Osuga Y, Fujii T. Recent advances in targeting DNA repair pathways for the treatment of ovarian cancer and their clinical relevance. Int J Clin Oncol. 2017;22:611–8.
10. Ledermann JA, Drew Y, Kristeleit RS. Homologous recombination deficiency and ovarian cancer. Eur J Cancer. 2016;60:49–58.
11. Lin KY, Kraus WL. PARP inhibitors for cancer therapy. Cell. 2017;169:183.
12. Robson M, Im SA, Senkus E, Xu B, Domchek SM, Masuda N, Delaloge S, Li W, Tung N, Armstrong A, Wu W, Goessl C, Runswick S, Conte P. Olaparib for metastatic breast cancer in patients with a Germline BRCA mutation. N Engl J Med. 2017;377:523–33.
13. Golan T, Hammel P, Reni M, Van Cutsem E, Macarulla T, Hall MJ, Park JO, Hochhauser D, Arnold D, Oh DY, Reinacher-Schick A, Tortora G, Algテシl H, O'Reilly EM, McGuinness D, Cui KY, Schlienger K, Locker GY, Kindler HL. Maintenance Olaparib for Germline BRCA-mutated metastatic pancreatic cancer. N Engl J Med. 2019;381:317–27.
14. de Bono J, Mateo J, Fizazi K, Saad F, Shore N, Sandhu S, Chi KN, Sartor O, , Agarwal N, Olmos D, Thiery-Vuillemin A, Twardowski P, Mehra N, Goessl C, Kang J, Burgents J, Wu W, Kohlmann A, Adelman CA, Hussain M. Olaparib for metastatic castration-resistant prostate cancer. N Engl J Med 382:2091–2102, 2020.
15. Witjes VM, van Bommel MHD, Ligtenberg MJL, Vos JR, Mourits MJE, Ausems MGEM, de Hullu JA, Bosse T, Hoogerbrugge N. Probability of detecting germline BRCA1/2 pathogenic variants in histological subtypes of ovarian carcinoma. A meta-analysis. Gynecol Oncol. 2022;164:221–30.
16. Telli ML, Timms KM, Reid J, Hennessy B, Mills GB, Jensen KC, Szallasi Z, Barry WT, Winer EP, Tung NM, Isakoff SJ, Ryan PD, Greene-Colozzi A, Gutin A, Sangale Z, Iliev D, Neff C, Abkevich V, Jones JT, Lanchbury JS, Hartman AR, Garber JE, Ford JM, Silver DP, Richardson AL. Homologous recombination deficiency (HRD) score predicts response to platinum-containing Neoadjuvant chemotherapy in patients with triple-negative breast cancer. Clin Cancer Res. 2016;22:3764–73.
17. Hodgson DR, Dougherty BA, Lai Z, Fielding A, Grinsted L, Spencer S, O'Connor MJ, Ho TW, Robertson JD, Lanchbury JS, Timms KM, Gutin A, Orr M, Jones H, Gilks B, Womack C, Gourley C, Ledermann J, Barrett JC. Candidate biomarkers of PARP inhibitor sensitivity in ovarian cancer beyond the BRCA genes. Br J Cancer. 2018;119:1401–9.
18. Kanchi KL, Johnson KJ, Lu C, McLellan MD, Leiserson MD, Wendl MC, Zhang Q, Koboldt DC, Xie M, Kandoth C, McMichael JF, Wyczalkowski MA, Larson DE, Schmidt HK, Miller CA, Fulton RS, Spellman PT, Mardis ER, Druley TE, Graubert TA, Goodfellow PJ, Raphael BJ, Wilson RK, Ding L. Integrated analysis of germline and somatic variants in ovarian cancer. Nat Commun. 2014;5:3156.
19. Sánchez-Lorenzo L, Salas-Benito D, Villamayor J, Patiño-García A, González-Martín A. The BRCA gene in epithelial ovarian cancer. Cancers (Basel). 2022;14:1235.
20. Konstantinopoulos PA, Norquist B, Lacchetti C, Armstrong D, Grisham RN, Goodfellow PJ, Kohn EC, Levine DA, Liu JF, Lu KH, Sparacio D, Annunziata CM. Germline and somatic tumor testing in epithelial ovarian cancer: ASCO guideline. J Clin Oncol. 2020;38:1222–45.
21. Hodgson DR, Brown JS, Dearden SP, Lai Z, Elks CE, Milenkova T, Dougherty BA, Lanchbury JS, Perry M, Timms KM, Harrington EA, Barrett JC, Leary A, Pujade-Lauraine E. Concordance of BRCA mutation detection in tumor versus blood, and frequency of bi-allelic loss of BRCA in tumors from patients in the phase III SOLO2 trial. Gynecol Oncol. 2021;163:563–8.
22. Ebi H, Bando H. Precision Oncology and the Universal Health Coverage System in Japan. JCO Precis Oncol. 2019;3:1–12.

23. Mukai Y, Ueno H. Establishment and implementation of cancer genomic medicine in Japan. Cancer Sci. 2021;112:970–7.
24. Takeda M, Takahama T, Sakai K, Shimizu S, Watanabe S, Kawakami H, Tanaka K, Sato C, Hayashi H, Nonagase Y, Yonesaka K, Takegawa N, Okuno T, Yoshida T, Fumita S, Suzuki S, Haratani K, Saigoh K, Ito A, Mitsudomi T, Handa H, Fukuoka K, Nakagawa K, Nishio K. Clinical application of the FoundationOne CDx assay to therapeutic decision-making for patients with advanced solid tumors. Oncologist. 2021;26:e588–96.
25. Sunami K, Ichikawa H, Kubo T, Kato M, Fujiwara Y, Shimomura A, Koyama T, Kakishima H, Kitami M, Matsushita H, Furukawa E, Narushima D, Nagai M, Taniguchi H, Motoi N, Sekine S, Maeshima A, Mori T, Watanabe R, Yoshida M, Yoshida A, Yoshida H, Satomi K, Sukeda A, Hashimoto T, Shimizu T, Iwasa S, Yonemori K, Kato K, Morizane C, Ogawa C, Tanabe N, Sugano K, Hiraoka N, Tamura K, Yoshida T, Fujiwara Y, Ochiai A, Yamamoto N, Kohno T. Feasibility and utility of a panel testing for 114 cancer-associated genes in a clinical setting: A hospital-based study. Cancer Sci. 2019;110:1480–90.
26. Yoshii Y, Okazaki S, Takeda M. Current status of next-generation sequencing-based cancer genome profiling tests in japan and prospects for liquid biopsy. Life (Basel). 2021;11:796.
27. Naito Y, Aburatani H, Amano T, Baba E, Furukawa T, Hayashida T, Hiyama E, Ikeda S, Kanai M, Kato M, Kinoshita I, Kiyota N, Kohno T, Kohsaka S, Komine K, Matsumura I, Miura Y, Nakamura Y, Natsume A, Nishio K, Oda K, Oda N, Okita N, Oseto K, Sunami K, Takahashi H, Takeda M, Tashiro S, Toyooka S, Ueno H, Yachida S, Yoshino T, Tsuchihara K. Clinical practice guidance for next-generation sequencing in cancer diagnosis and treatment (edition 2.1). Int J Clin Oncol. 2021;26:233–83.
28. Minamoto A, Yamada T, Shimada S, Kinoshita I, Aoki Y, Oda K, Ueki A, Higashigawa S, Morikawa M, Sato Y, Hirasawa A, Ogawa M, Kondo T, Yoshioka M, Kanai M, Muto M, Kosugi S. Current status and issues related to secondary findings in the first public insurance covered tumor genomic profiling in Japan: multi-site questionnaire survey. J Hum Genet. 2022:1–7.
29. Kohsaka S, Tatsuno K, Ueno T, Nagano M, Shinozaki-Ushiku A, Ushiku T, Takai D, Ikegami M, Kobayashi H, Kage H, Ando M, Hata K, Ueda H, Yamamoto S, Kojima S, Oseto K, Akaike K, Suehara Y, Hayashi T, Saito T, Takahashi F, Takahashi K, Takamochi K, Suzuki K, Nagayama S, Oda Y, Mimori K, Ishihara S, Yatomi Y, Nagase T, Nakajima J, Tanaka S, Fukayama M, Oda K, Nangaku M, Miyazono K, Miyagawa K, Aburatani H, Mano H. Comprehensive assay for the molecular profiling of cancer by target enrichment from formalin-fixed paraffin-embedded specimens. Cancer Sci. 2019;110:1464–79.
30. Kage H, Kohsaka S, Tatsuno K, Ueno T, Ikegami M, Zokumasu K, Shinozaki-Ushiku A, Nagai S, Aburatani H, Mano H, Oda K. Tumor mutational burden measurement using comprehensive genomic profiling assay. Jpn J Clin Oncol. 2022;52(8):917–21.
31. Pestinger V, Smith M, Sillo T, Findlay JM, Laes JF, Martin G, Middleton G, Taniere P, Beggs AD. Use of an integrated pan-cancer oncology enrichment next-generation sequencing assay to measure tumour mutational burden and detect clinically actionable variants. Mol Diagn Ther. 2020;24:339–49.
32. Woodhouse R, Li M, Hughes J, Delfosse D, Skoletsky J, Ma P, Meng W, Dewal N, Milbury C, Clark T, Donahue A, Stover D, Kennedy M, Dacpano-Komansky J, Burns C, Vietz C, Alexander B, Hegde P, Dennis L. Clinical and analytical validation of FoundationOne liquid CDx, a novel 324-gene cfDNA-based comprehensive genomic profiling assay for cancers of solid tumor origin. PLoS One. 2020;15:e0237802.
33. Nakamura Y, Taniguchi H, Ikeda M, Bando H, Kato K, Morizane C, Esaki T, Komatsu Y, Kawamoto Y, Takahashi N, Ueno M, Kagawa Y, Nishina T, Kato T, Yamamoto Y, Furuse J, Denda T, Kawakami H, Oki E, Nakajima T, Nishida N, Yamaguchi K, Yasui H, Goto M, Matsuhashi N, Ohtsubo K, Yamazaki K, Tsuji A, Okamoto W, Tsuchihara K, Yamanaka T, Miki I, Sakamoto Y, Ichiki H, Hata M, Yamashita R, Ohtsu A, Odegaard JI, Yoshino T. Clinical utility of circulating tumor DNA sequencing in advanced gastrointestinal cancer: SCRUM-Japan GI-SCREEN and GOZILA studies. Nat Med. 2020;26:1859–64.

34. Slavin TP, Banks KC, Chudova D, Oxnard GR, Odegaard JI, Nagy RJ, Tsang KWK, Neuhausen SL, Gray SW, Cristofanilli M, Rodriguez AA, Bardia A, Leyland-Jones B, Janicek MF, Lilly M, Sonpavde G, Lee CE, Lanman RB, Meric-Bernstam F, Kurzrock R, Weitzel JN. Identification of incidental Germline mutations in patients with advanced solid tumors who underwent cell-free circulating tumor DNA sequencing. J Clin Oncol. 2018;36:JCO1800328.
35. Ledermann J, Harter P, Gourley C, Friedlander M, Vergote I, Rustin G, Scott C, Meier W, Shapira-Frommer R, Safra T, Matei D, Macpherson E, Watkins C, Carmichael J, Matulonis U. Olaparib maintenance therapy in platinum-sensitive relapsed ovarian cancer. N Engl J Med. 2012;366:1382–92.
36. Mirza MR, Monk BJ, Herrstedt J, Oza AM, Mahner S, Redondo A, Fabbro M, Ledermann JA, Lorusso D, Vergote I, Ben-Baruch NE, Marth C, Monk BJ, Herrstedt J, Oza AM, Mahner S, Redondo A, Fabbro M, Ledermann JA, Lorusso D, Vergote I, Ben-Baruch NE. Niraparib maintenance therapy in platinum-sensitive, recurrent ovarian cancer. N Engl J Med. 2016;375:2154–64.
37. Pujade-Lauraine E, Ledermann JA, Selle F, Gebski V, Penson RT, Oza AM, Korach J, Huzarski T, Poveda A, Pignata S, Friedlander M, Colombo N, Harter P, Fujiwara K, Ray-Coquard I, Banerjee S, Liu J, Lowe ES, Bloomfield R, Pautier P. Olaparib tablets as maintenance therapy in patients with platinum-sensitive, relapsed ovarian cancer and a BRCA1/2 mutation (SOLO2/ENGOT-Ov21): a double-blind, randomised, placebo-controlled, phase 3 trial. Lancet Oncol. 2017;18:1274–84.
38. Moore K, Colombo N, Scambia G, Kim BG, Oaknin A, Friedlander M, Lisyanskaya A, Floquet A, Leary A, Sonke GS, Gourley C, Banerjee S, Oza A, Gonzт̄。lez-Martт̄ın A, Aghajanian C, Bradley W, Mathews C, Liu J, Lowe ES, Bloomfield R, Di Silvestro P. Maintenance Olaparib in patients with newly diagnosed advanced ovarian cancer. N Engl J Med 379:2495–2505, 2018.
39. Moore KN, Secord AA, Geller MA, Miller DS, Cloven N, Fleming GF, Wahner Hendrickson AE, Azodi M, DiSilvestro P, Oza AM, Cristea M, Berek JS, Chan JK, Rimel BJ, Matei DE, Li Y, Sun K, Luptakova K, Matulonis UA, Monk BJ. Niraparib monotherapy for late-line treatment of ovarian cancer (QUADRA): a multicentre, open-label, single-arm, phase 2 trial. Lancet Oncol. 2019;20:636–48.
40. Ray-Coquard I, Pautier P, Pignata S, Pт̄ϑrol D, Gonzт̄lez-Martт̄ın A, Berger R, Fujiwara K, Vergote I, Colombo N, Mт̄ˎenpт̄ˎт̄ˎ J, Selle F, Sehouli J, Lorusso D, Guerra Alт̄ıa EM, Reinthaller A, Nagao S, Lefeuvre-Plesse C, Canzler U, Scambia G, Lortholary A, Marmт̄ϑ F, Combe P, de Gregorio N, Rodrigues M, Buderath P, Dubot C, Burges A, You B, Pujade-Lauraine E, Harter P Olaparib plus Bevacizumab as First-Line Maintenance in Ovarian Cancer N Engl J Med 381:2416–2428, 2019.
41. Gonzalez-Martin A, Pothuri B, Vergote I, DePont CR, Graybill W, Mirza MR, McCormick C, Lorusso D, Hoskins P, Freyer G, Baumann K, Jardon K, Redondo A, Moore RG, Vulsteke C, O'Cearbhaill RE, Lund B, Backes F, Barretina-Ginesta P, Haggerty AF, Rubio-Pт̄ϑrez MJ, Shahin MS, Mangili G, Bradley WH, Bruchim I, Sun K, Malinowska IA, Li Y, Gupta D, Monk BJ. Niraparib in patients with newly diagnosed advanced ovarian cancer. N Engl J Med. 2019;381:2391–402.
42. PujadeLauraine E, Selle F, Scambia G, Asselain B, Marmé F, Lindemann K, Colombo N, Madry R, Glasspool RM, Dubot C, Oaknin A, Zamagni C, Heitz F, Gladieff L, RubioPérez MJ, Scollo P, Blakeley C, Shaw B, RayCoquard IL, A.. Maintenance olaparib rechallenge in patients (pts) with ovarian carcinoma (OC) previously treated with a PARP inhibitor (PARPi): phase IIIb OReO/ENGOT Ov-38 trial. Ann Oncol. 2021;2021(32):S1283–346. https://doi.org/10.1016/annonc/annonc741.
43. OPINION primary analysis. Poveda A, Lheureux S, Colombo N, Cibula D, Lindemann K, Weberpals J, Bjurberg M, Oaknin A, Sikorska M, Gonzт̄。lez-Martт̄ın A, Madry R, Pт̄ϑrez MJR, Ledermann J, Davidson R, Blakeley C, Bennett J, Barnicle A, †?kof E. Olaparib maintenance monotherapy in platinum-sensitive relapsed ovarian cancer patients without a germline BRCA1/BRCA2 mutation. Gynecol Oncol. 2022;164:498–504.

44. Abbotts R, Wilson DM 3rd. Coordination of DNA single strand break repair. Free Radic Biol Med. 2017;107:228–44.
45. Dianov GL, Hübscher U. Mammalian base excision repair: the forgotten archangel. Nucleic Acids Res. 2013;41:3483–90.
46. Hosoya N, Miyagawa K. Targeting DNA damage response in cancer therapy. Cancer Sci. 2014;105:370–88.
47. Valikhani M, Rahimian E, Ahmadi SE, Chegeni R, Safa M. Involvement of classic and alternative non-homologous end joining pathways in hematologic malignancies: targeting strategies for treatment. Exp Hematol Oncol. 2021;10:51.
48. Caracciolo D, Riillo C, Di Martino MT, Tagliaferri P, Tassone P. Alternative non-homologous end-joining: error-prone DNA repair as cancer's achilles' heel. Cancers (Basel). 2021;13:1392.
49. Ramsden DA, Carvajal-Garcia J, Gupta GP. Mechanism, cellular functions and cancer roles of polymerase-theta-mediated DNA end joining. Nat Rev Mol Cell Biol. 2022;23:125–40.
50. Bhargava R, Onyango DO, Stark JM. Regulation of single-Strand annealing and its role in genome maintenance. Trends Genet. 2016;32:566–75.
51. Brown JS, O'Carrigan B, Jackson SP, Yap TA. Targeting DNA repair in cancer: beyond PARP inhibitors. Cancer Discov. 2017;7:20–37.
52. Burdak-Rothkamm S, Rothkamm K. DNA damage repair deficiency and synthetic lethality for cancer treatment. Trends Mol Med. 2021;27:91–2.
53. Zong C, Zhu T, He J, Huang R, Jia R, Shen J. PARP mediated DNA damage response, genomic stability and immune responses. Int J Cancer. 2022;150:1745–59.
54. Michl J, Zimmer J, Tarsounas M. Interplay between Fanconi anemia and homologous recombination pathways in genome integrity. EMBO J. 2016;35:909–23.
55. Kim H, D'Andrea AD. Regulation of DNA cross-link repair by the Fanconi anemia/BRCA pathway. Genes Dev. 2012;26:1393–408.
56. Haynes B, Saadat N, Myung B, Shekhar MP. Crosstalk between translesion synthesis, Fanconi anemia network, and homologous recombination repair pathways in interstrand DNA crosslink repair and development of chemoresistance. Mutat Res Rev Mutat Res. 2015;763:258–66.
57. Ceccaldi R, Sarangi P, D'Andrea AD. The Fanconi anaemia pathway: new players and new functions. Nat Rev Mol Cell Biol. 2016;17:337–49.
58. Hoogenboom WS, Boonen RACM, Knipscheer P. The role of SLX4 and its associated nucleases in DNA interstrand crosslink repair. Nucleic Acids Res. 2019;47:2377–88.
59. Maiorano D, El Etri J, Franchet C, Hoffmann JS. Translesion synthesis or repair by specialized DNA polymerases limits excessive genomic instability upon replication stress. Int J Mol Sci. 2021;22:3924.
60. Murai J, Huang SY, Das BB, Renaud A, Zhang Y, Doroshow JH, Ji J, Takeda S, Pommier Y. Trapping of PARP1 and PARP2 by clinical PARP inhibitors. Cancer Res. 2012;72:5588–99.
61. Min A, Im SA. PARP inhibitors as therapeutics: beyond modulation of PARylation. Cancers (Basel). 2020;12:394.
62. Saha LK, , Murai Y, Saha S, Jo U, Tsuda M, Takeda S, Pommier Y. Replication-dependent cytotoxicity and Spartan-mediated repair of trapped PARP1-DNA complexes. Nucleic Acids Res 49:10493–10506, 2021.
63. Loman N, Johannsson O, Kristoffersson U, Olsson H, Borg A. Family history of breast and ovarian cancers and BRCA1 and BRCA2 mutations in a population-based series of early-onset breast cancer. J Natl Cancer Inst. 2001;93:1215–23.
64. Radice P. Mutations of BRCA genes in hereditary breast and ovarian cancer. J Exp Clin Cancer Res. 2002;21:9–12.
65. Kim G, Ison G, McKee AE, Zhang H, Tang S, Gwise T, Sridhara R, Lee E, Tzou A, Philip R, Chiu HJ, Ricks TK, Palmby T, Russell AM, Ladouceur G, Pfuma E, Li H, Zhao L, Liu Q, Venugopal R, Ibrahim A, Pazdur R. FDA approval summary: Olaparib Monotherapy in patients with deleterious Germline BRCA-mutated advanced ovarian cancer treated with three or more lines of chemotherapy. Clin Cancer Res. 2015;21:4257–61.

66. Kaufman B, Shapira-Frommer R, Schmutzler RK, Audeh MW, Friedlander M, Balmaテアa J, Mitchell G, Fried G, Stemmer SM, Hubert A, Rosengarten O, Steiner M, Loman N, Bowen K, Fielding A, Domchek SM. Olaparib monotherapy in patients with advanced cancer and a germline BRCA1/2 mutation. J Clin Oncol. 2015;33:244–50.
67. Coleman RL, Oza AM, Lorusso D, Aghajanian C, Oaknin A, Dean A, Colombo N, Weberpals JI, Clamp A, Scambia G, Leary A, Holloway RW, Gancedo MA, Fong PC, Goh JC, O'Malley DM, Armstrong DK, Garcia-Donas J, Swisher EM, Floquet A, Konecny GE, McNeish IA, Scott CL, Cameron T, Maloney L, Isaacson J, Goble S, Grace C, Harding TC, Raponi M, Sun J, Lin KK, Giordano H, Ledermann JA. Rucaparib maintenance treatment for recurrent ovarian carcinoma after response to platinum therapy (ARIEL3): a randomised, double-blind, placebo-controlled, phase 3 trial. Lancet. 2017;390:1949–61.
68. Coleman RL, Fleming GF, Brady MF, Swisher EM, Steffensen KD, Friedlander M, Okamoto A, Moore KN, Efrat BenBaruch N, Werner TL, Cloven NG, Oaknin A, DiSilvestro PA, Morgan MA, Nam JH, Leath CA 3rd, Nicum S, Hagemann AR, Littell RD, Cella D, Baron-Hay S, Garcia-Donas J, Mizuno M, Bell-McGuinn K, Sullivan DM, Bach BA, Bhattacharya S, Ratajczak CK, Ansell PJ, Dinh MH, Aghajanian C, Bookman MA. Veliparib with first-line chemotherapy and as maintenance therapy in ovarian cancer. N Engl J Med. 2019;381:2403–15.
69. Dougherty BA, Lai Z, Hodgson DR, Orr MCM, Hawryluk M, Sun J, Yelensky R, Spencer SK, Robertson JD, Ho TW, Fielding A, Ledermann JA, Barrett JC. Biological and clinical evidence for somatic mutations in BRCA1 and BRCA2 as predictive markers for olaparib response in high-grade serous ovarian cancers in the maintenance setting. Oncotarget. 2017;8:43653–61.
70. Mateo J, Lord CJ, Serra V, Tutt A, Balmaテアa J, Castroviejo-Bermejo M, Cruz C, Oaknin A, Kaye SB, de Bono JS. A decade of clinical development of PARP inhibitors in perspective. Ann Oncol. 2019;30:1437–47.
71. Marquard AM, Eklund AC, Joshi T, Krzystanek M, Favero F, Wang ZC, Richardson AL, Silver DP, Szallasi Z, Birkbak NJ. Pan-cancer analysis of genomic scar signatures associated with homologous recombination deficiency suggests novel indications for existing cancer drugs. Biomark Res. 2015;3:9.
72. Swisher EM, Lin KK, Oza AM, Scott CL, Giordano H, Sun J, Konecny GE, Coleman RL, Tinker AV, O'Malley DM, Kristeleit RS, Ma L, Bell-McGuinn KM, Brenton JD, Cragun JM, Oaknin A, Ray-Coquard I, Harrell MI, Mann E, Kaufmann SH, Floquet A, Leary A, Harding TC, Goble S, Maloney L, Isaacson J, Allen AR, Rolfe L, Yelensky R, Raponi M, McNeish IA. Rucaparib in relapsed, platinum-sensitive high-grade ovarian carcinoma (ARIEL2 part 1): an international, multicentre, open-label, phase 2 trial. Lancet Oncol. 2017;18:75–87.
73. Norquist BM, Harrell MI, Brady MF, Walsh T, Lee MK, Gulsuner S, Bernards SS, Casadei S, Yi Q, Burger RA, Chan JK, Davidson SA, Mannel RS, DiSilvestro PA, Lankes HA, Ramirez NC, King MC, Swisher EM, Birrer MJ. Inherited mutations in women with ovarian carcinoma. JAMA Oncol. 2016;2:482–90.
74. Couch FJ, Hart SN, Sharma P, Toland AE, Wang X, Miron P, Olson JE, Godwin AK, Pankratz VS, Olswold C, Slettedahl S, Hallberg E, Guidugli L, Davila JI, Beckmann MW, Janni W, Rack B, Ekici AB, Slamon DJ, Konstantopoulou I, Fostira F, Vratimos A, Fountzilas G, Pelttari LM, Tapper WJ, Durcan L, Cross SS, Pilarski R, Shapiro CL, Klemp J, Yao S, Garber J, Cox A, Brauch H, Ambrosone C, Nevanlinna H, Yannoukakos D, Slager SL, Vachon CM, Eccles DM, Fasching PA. Inherited mutations in 17 breast cancer susceptibility genes among a large triple-negative breast cancer cohort unselected for family history of breast cancer. J Clin Oncol. 2015;33:304–11.
75. Matsumoto K, Nishimura M, Onoe T, Sakai H, Urakawa Y, Onda T, Yaegashi N. PARP inhibitors for BRCA wild type ovarian cancer; gene alterations, homologous recombination deficiency and combination therapy. Jpn J Clin Oncol. 2019;49:703–7.
76. Norquist BM, Brady MF, Harrell MI, Walsh T, Lee MK, Gulsuner S, Bernards SS, Casadei S, Burger RA, Tewari KS, Backes F, Mannel RS, Glaser G, Bailey C, Rubin S, Soper J, Lankes HA, Ramirez NC, King MC, Birrer MJ, Swisher EM. Mutations in homologous recombination

genes and outcomes in ovarian carcinoma patients in GOG 218: an NRG oncology/gynecologic oncology group study. Clin Cancer Res. 2018;24:777–83.
77. Hansmann T, Pliushch G, Leubner M, Kroll P, Endt D, Gehrig A, Preisler-Adams S, Wieacker P, Haaf T. Constitutive promoter methylation of BRCA1 and RAD51C in patients with familial ovarian cancer and early-onset sporadic breast cancer. Hum Mol Genet. 2012;21:4669–79.
78. Bernards SS, Pennington KP, Harrell MI, Agnew KJ, Garcia RL, Norquist BM, Swisher EM. Clinical characteristics and outcomes of patients with BRCA1 or RAD51C methylated versus mutated ovarian carcinoma. Gynecol Oncol. 2018;148:281–5.
79. Chiang JW, Karlan BY, Cass L, Baldwin RL. BRCA1 promoter methylation predicts adverse ovarian cancer prognosis. Gynecol Oncol. 2006;101:403–10.
80. Lin KK, Harrell MI, Oza AM, Oaknin A, Ray-Coquard I, Tinker AV, Helman E, Radke MR, Say C, Vo LT, Mann E, Isaacson JD, Maloney L, O'Malley DM, Chambers SK, Kaufmann SH, Scott CL, Konecny GE, Coleman RL, Sun JX, Giordano H, Brenton JD, Harding TC, McNeish IA, Swisher EM. BRCA reversion mutations in circulating tumor DNA predict primary and acquired resistance to the PARP inhibitor Rucaparib in high-grade ovarian carcinoma. Cancer Discov. 2019;9:210–9.
81. Wan JCM, Massie C, Garcia-Corbacho J, Mouliere F, Brenton JD, Caldas C, Pacey S, Baird R, Rosenfeld N. Liquid biopsies come of age: towards implementation of circulating tumour DNA. Nat Rev Cancer. 2017;17:223–38.
82. Weigelt B, Comino-Mテウndez I, de Bruijn I, Tian L, Meisel JL, Garcia-Murillas I, Fribbens C, Cutts R, Martelotto LG, Ng CKY, Lim RS, Selenica P, Piscuoglio S, Aghajanian C, Norton L, Murali R, Hyman DM, Borsu L, Arcila ME, Konner J, Reis-Filho JS, Greenberg RA, Robson ME, Turner NC. Diverse BRCA1 and BRCA2 reversion mutations in circulating cell-free DNA of therapy-resistant breast or ovarian cancer. Clin Cancer Res. 2017;23:6708–20.
83. Tobalina L, Armenia J, Irving E, O'Connor MJ, Forment JV. A meta-analysis of reversion mutations in BRCA genes identifies signatures of DNA end-joining repair mechanisms driving therapy resistance. Ann Oncol. 2021;32:103–12.

# Chapter 6
# Personalized Treatment in Immunotherapy for Gynecologic Cancer

**Junzo Hamanishi**

**Abstract** Recent cancer treatments have entered a new era with novel types of immunotherapies. In particular, immune checkpoint signals mediated by the immunosuppressive cofactors programmed cell death 1 (PD-1) and PD-1 ligand 1 (PD-L1), are the most promising targets for new cancer treatments.

Several clinical trials of various types of gynecologic cancers have been completed and revealed a modest antitumor effect with monotherapy with immune checkpoint inhibitors (ICIs; anti-PD-1 antibody and/or anti-PD-L1 antibody). However, genetic and/or molecular biomarker-selected endometrial cancer and cervical cancers are more promising for treatment with ICIs. Some ICIs have been approved by the FDA and the combination of ICIs with other agents has yielded good results in trials for these cancers. Therefore, the selection of patients who would benefit from ICI immunotherapy is quite important.

**Keywords** Immune checkpoint inhibition · PD-1 · PD-L1 · MSI · TMB

## 6.1 Introduction

In the last decade, cancer treatment has been revolutionized by new types of immunotherapies, mainly immune checkpoint inhibitors (ICIs) such as anti-programmed cell death 1 (PD-1) antibodies and/or anti-PD ligand 1 (PD-L1) antibodies (Table 6.1), which have become standard treatments for several advanced solid tumors [1, 2] (Table 6.2). Based on the mechanism of action of these agents, several biomarkers to measure the response to treatment have been investigated in

J. Hamanishi (✉)
Department of Gynecology and Obstetrics, Kyoto University Graduate School of Medicine, Kyoto, Japan
e-mail: jnkhmns@kuhp.kyoto-u.ac.jp


M. Mandai (ed.), *Personalization in Gynecologic Oncology*, Comprehensive Gynecology and Obstetrics, https://doi.org/10.1007/978-981-19-4711-7_6

**Table 6.1** Immune checkpoint inhibitors (PD-1 signal inhibitors) in gynecologic cancers

| Target | Agent | Brand name | Company |
|---|---|---|---|
| PD1 | Nivolumab | Opdivo | Bristol-Meyers Squibb/Ono |
| | Pembrolizumab | Keytruda | MSD |
| | Dostarlimab-gxly | Jemperli | GSK |
| | Cemiplimab-rwlc | Libtayo | Sanofi |
| | Balstilimab | | Agenus |
| PD-L1 | Atezolizumab | Tecentriq | Roche |
| | Durvalumab | Imfinzi | AstraZeneca |
| | Avelumab | BAVENCIO | Pfizer |

**Table 6.2** FDA approved PD-1 signal inhibitors in gynecologic cancers

| Tumor type | Biomaker | Agent | Company |
|---|---|---|---|
| Cervical cancer | PD-L1 | Pembrolizumab | MSD |
| | PD-L1 | Pembrolizumab±TC+Bmab | MSD |
| Endometrial cancer | MSS | Pembrolizumab+lenvatinib | MSD/Eisai |
| | dMMR/MSI-High | Dostarlimab-gxly | GSK |
| Solid tumor | dMMR/MSI-High | Dostarlimab-gxly | |
| | | Pembrolizumab | MSD |
| | TMB-High | Pembrolizumab | |

*MSI-High, microsatellite instability-high; MSS, microsatellite stable; MMRd, mismatch repair deficiency; TC±Bmab, paclitxel±bevacizumab

clinical trials and have led to the approval of ICI-based treatments [3, 4]. More recently, some clinical trials using ICI monotherapy have demonstrated promising antitumor effects for gynecologic cancers such as mismatch repair deficient (MMRd) or microsatellite instability (MSI)-high cases of endometrial cancers and PD-L1-expressing cervical cancer [5]. However, gynecologic cancers, particularly ovarian cancer, represent a heterogenous subgroup of histologies, and thus their responses to ICIs cannot be fully predicted using known biomarkers. Therefore, the optimal biomarkers for specific subtypes of patients with cancer are urgently required [4].

Conversely, combination immunotherapies with other antitumor therapies such as chemotherapy, targeted therapy, radiotherapy, or other immunotherapies have been expected to enhance the antitumor effect of ICIs in gynecologic cancers, and some of these have demonstrated promising synergistic effects.

This chapter highlights the mechanism of action of ICIs and recent clinical trials of ICIs used to treat gynecologic cancers, including specific molecular- or genetic-based personalized immunotherapies (Fig. 6.1).

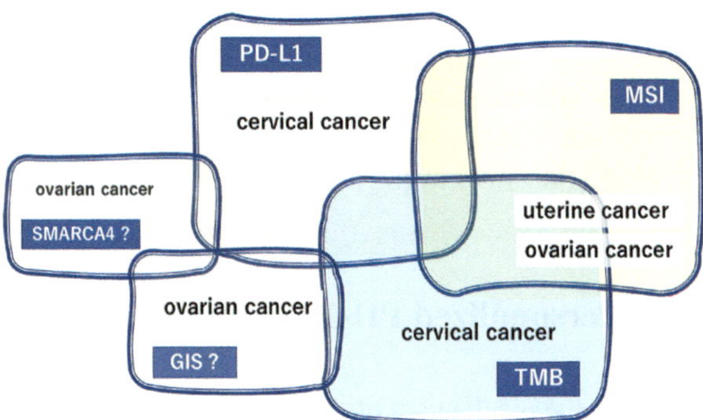


**Fig. 6.1** Precision biomarkers in gynecologic cancers. MSI, microsatellite instability; TMB, tumor mutational burden; GIS, genomic instability score; SMARCA4, SWI/SNF-related, matrix-associated, actin-dependent regulator of chromatin, subfamily A, member 4

## 6.2 PD-1 Signal

PD-1 (CD279) is an immunosuppressive co-inhibitory molecule that belongs to the CD28 family of receptors on T cells. PD-1 was discovered in 1992 and is known to be an induced molecule on T cells undergoing apoptosis [6]. Additional studies demonstrated PD-1 expression on mature hematopoietic cells such as T and B cells, as well as monocytes, following activation [7]. The cognate ligands for PD-1 are the B7-family molecules, PD-L1 (CD274, B7-H1), and PD-L2 (CD273, B7-H2). PD-L1 is expressed in human tonsils, placenta, monocytes, and lungs, where it plays a role in immune tolerance. PD-L2 is mainly expressed in dendritic cells (DCs) under normal physiological conditions [8]. The PD-1/PD-L1/L2 signaling pathway has been shown to control excessive autoimmune and inflammatory responses. This pathway plays a key role in immune homeostasis, together with the B7–1/2/CTLA-4 signaling pathway described above [9]. The CTLA-4 signaling pathway is primarily involved in the process of antigen presentation in lymph nodes, whereas the key role of the PD-1 signaling pathway is to suppress the immune response to target cancer cells in peripheral tissue.

## 6.3 PD-1 Signal Inhibitors

Several clinical trials have utilized humanized anti-PD-1 antibodies (pembrolizumab, nivolumab, dostarlimab-gxly, cemiplimab-rwlc, and balstilimab) and anti-PD-L1 antibodies (atezolizumab, durvalumab, avelumab) for solid tumors as PD-1 signal inhibitors (Table 6.1). Subsequently, some antibodies have been approved by the FDA for gynecologic cancers (Table 6.2). Pembrolizumab, an anti-PD-1 antibody, was first approved for melanoma in 2014 and dostarlimab-gxly, another anti-PD-1 antibody, has been approved for mismatch repair deficient (MMRd) recurrent or advanced solid tumors. To date, at least 200 clinical studies have been carried out using some type of PD-1 signal inhibitors in gynecologic cancers (ClinicalTrials.gov) [10].

## 6.4 Personalized PD-1 Signal Inhibitors

Based on the mechanism of action of PD-1 signal inhibitors, several biomarkers for the response to treatment have been investigated in various clinical trials, including gynecologic cancers, and this has led to the approval of PD-1 signal inhibitors. Recent clinical trials of PD-1 signal inhibitors for gynecological cancer have demonstrated promising results in endometrial cancers and cervical cancers. Because gynecological cancers represent a heterogeneous group of tumors, the optimal biomarkers for a specific type of cancer have not yet been fully determined; some biological biomarkers known to be useful for the treatment of gynecologic cancers are shown in Fig. 6.1.

## 6.5 PD-L1 Expression

PD-L1 protein expression is a predictive biomarker of the efficacy of PD-1 inhibitors in several types of cancer including cervical cancer, but not in endometrial cancer and ovarian cancer, as demonstrated in previous immunotherapeutic clinical trials [11, 12]. PD-L1 gene amplification has been reported in 0.7% of 100,000 cases of more than 100 types of solid tumors, and specifically in 2.7% (10/374) cases of cervical cancer. Furthermore, although the number of cases is small, the response rate (RR) to PD-1 pathway inhibitors was 66.7% (6 responses) in 9 solid tumors with PD-L1 gene amplification [10], and the same has been reported in multiple cancer types including lung cancer and breast cancer. Therefore, PD-1 pathway inhibitors are often recommended for cancers with high PD-L1 gene expression including cervical cancers, and not for endometrial and ovarian cancers.

In the KEYNOTE-158 trial of pembrolizumab in previously treated recurrent cervical cancer, the RR was 14.6% in patients with high PD-L1 expression and 0% in patients with low expression. The FDA approved pembrolizumab for PD-L1–positive recurrent cervical cancer in 2018 [13]. The KEYNOTE-826 trial studied the use of pembrolizumab in untreated metastatic/advanced cervical cancer and combined standard chemotherapy with bevacizumab with or without pembrolizumab. This study found that both progression-free survival and overall survival improved in the pembrolizumab group, and following this, the FDA approved pembrolizumab as a first-line treatment for PD-L1 positive cervical cancer in 2022 [14]. Furthermore, in the randomized phase III EMPOWER-Cervical 1/GOG-3016/ENGOT-cx9 study with the novel anti-PD-1 antibody cemiplimab, cemiplimab demonstrated significantly longer overall survival compared to chemotherapy [11]. In the RaPiDS trial, another anti-PD-1 antibody, balstilimab, also demonstrated that the objective response rate (ORR) in PD-L1-positive patients was 20% and 8% in PD-L1 negative patients. The combination immunotherapy of balstilimab and the CTLA-4 inhibitor zalifrelimab resulted in an ORR of 27% and 11% in the PD-L1-positive and PD-L1-negative cohorts, respectively [15].

On the other hand, previous clinical trials of PD-1 signal inhibitors have shown no significant effect on PD-L1 expression in other gynecologic cancers [16].

## 6.6 Microsatellite Instability

Recent reports identified the frequency of genetic mutations derived from high microsatellite instability (MSI-High) with DNA mismatch repair deficiency (MMRd) in cancer cells as a candidate biomarker [17]. Many mutated antigens (called neoantigens) produced by MSI expressed on the surface of cancer cells are recognized by T cells and B cells as foreign antigens, either directly or through the APC system. Cancer cells exposed to IFN-γ released from activated T cells express PD-L1, thereby establishing an acquired immune resistance [18]. In this case, PD-1 signal inhibitors are more likely to be effective.

The frequency of MMRd /MSI-High in cancer varies by cancer type. In an analysis of 12,019 patients with MSI with 32 different types of cancer, MSI-High was identified in patients with 24 different carcinomas (2.2%). Endometrial cancer was the most common (17%) cancer with MSI-High, while ovarian cancer (3%) and cervical cancer were rare [5]. Therefore, MMRd/MSI-High endometrial cancer has become the focus of research as a good target for PD-1 signal inhibitors.

In the KEYNOTE-058 study of pembrolizumab in MMRd/MSI-High solid tumors, a high response rate was reported for MSI-H endometrial cancer in 28 of 49 patients (RR: 57%) and also for MMRd/MSI-High ovarian cancer in 5 of 15 patients (RR: 33%). The FDA has approved pembrolizumab for MMRd/MSI-High solid tumors across cancer types [19]. In addition to pembrolizumab, several other PD-1 pathway inhibitors such as dostarlimab-gxly (PD-1) and the antibodies avelumab (PD-L1) and durvalumab (PD-L1) have also been shown to be significantly more effective in patients with MMRd or MSI-High [20–22].

On the other hand, the KEYNOTE146 trial (Phase I/II), which investigated the efficacy of combination therapy of pembrolizumab with the multi-kinase inhibitor lenvatinib, showed a high RR of 57%, and was approved by the FDA for microsatellite stable (MSS)/mismatch repair proficient (MMRp) in 2019. The KEYNOTE-775/309 trial (phase III) demonstrated pembrolizumab in combination with lenvatinib prolonged overall survival, regardless of MMR abnormality [23].

## 6.7 Tumor Mutational Burden (TMB)

Recent comprehensive genetic mutation analysis of cancer tissues using next generation sequencing has revealed that, among the same cancer types, patients with high somatic gene mutations (tumor mutational burden-high: TMB-High) have high immunogenicity (immunoreactivity) due to the release of neoantigens, and the PD-1 pathway is induced by immune homeostatic reactions. In this situation, PD-1 pathway inhibitors reactivate the antitumor effect with increased immune cell infiltration into the tumor [24].

The KEYNOTE-158 trial of pembrolizumab in multiple solid tumors showed that the RR of the TMB-High group (TMB ≥10 mut/Mb, n = 102) was 29%, while that of the Non-TMB-High group ($n = 688$) was 6% [25]. In 2020, the FDA approved pembrolizumab for TMB-High solid tumors (≥10 mut/Mb), including gynecologic cancers.

The threshold of 10 mut/Mb of TMB-High is open to some debate. Recent research has demonstrated that not all types of TMB-High tumors such as brain tumors, unselected colon cancer, and esophageal cancer have demonstrated a favorable response to PD-1 signal inhibitors [26]. As for gynecological cancers, TMB-H predicts good responses in endometrial cancer but not in cervical cancer and ovarian cancer [27, 28].

## 6.8 Genomic Instability Score

Genomic Instability Score (GIS) is an algorithmic measurement of loss of heterozygosity, telomeric allelic imbalance, and large-scale state transitions by a next generation sequencing-based in vitro diagnostic test of tumor tissue specimens [29]. The results of the test are used to aid the identification of patients with ovarian cancer with positive homologous recombination deficiency status, who are eligible for treatment with poly (ADP-ribose) polymerase (PARP) inhibitors. It is thought that GIS-high will become a good biomarker of combination immunotherapy for ovarian cancer [30]. In the phase II study of the PARP inhibitor olaparib with the anti-PD-L1 antibody durvalumab (MEDIOLA doublet cohort) for germline BRCA-mutated platinum-sensitive relapsed (PSR) ovarian cancer, high ORR (72%) and disease control rates (81%) were observed with 19% of complete response (CR) cases [31]. Additionally, in the MEDIOLA triplet cohort, the triple combination immunotherapy of olaparib, durvalumab, and the anti-VEGF antibody bevacizumab for non-gBRCAm PSR ovarian cancer demonstrated an incredible antitumor effect in biomarker selected patients. Subgroup analysis revealed that a 100% RR (10 of 10 patients) was reported in GIS-positive cases (Foundation Medicine tumor analysis), while a 75% RR (6 of 8 patients) was reported in GIS-negative patients [32].

## 6.9 SMARCA4

SWI/SNF-Related, Matrix-Associated, Actin-Dependent Regulator of Chromatin, Subfamily A, Member 4 (SMARCA4) (also known as BRG1) is a subunit of the SWI/SNF chromatin remodeling complex, which regulates the expression of several genes [33]. Alterations in the SWI/SNF complex, and in particular, the loss of SMARCA4 expression, are well documented in several types of cancers including ovarian cancer. Small cell carcinoma of the ovary, hypercalcemic type (SCCOHT), is a rare, highly aggressive form of ovarian cancer seen primarily in younger patients, and has low survival rates for later-stage disease. Although low TMB with high intratumoral immune cell infiltration of SCCOHT would not predict responsiveness to an immune checkpoint blockade, a combination of PD-1 inhibitors with pembrolizumab have shown substantial and durable responses in selected patients [34].

## 6.10 Summary

Advances in clinical oncology and novel drug discoveries are playing a major role in personalized medicine in gynecologic cancers. The final goal of treatment for cancer is focused on patient specificity so that effective treatment is given to the right patient. The heterogeneity between gynecologic cancers among patients and within the same patient must be accounted for in personalized medicine. Therefore, genomic analyses with next generation sequencing and/or gene expression profiling using DNA microarrays along with bioinformatics will comprehensively reveal the diversity of the genome, epigenome, and expression profiles of gynecologic cancers that can be treated with immunotherapies. In the real-world clinical practice in medical oncology, we should perform reverse-translational research by using patients' samples such as tumor biopsies and/or blood samples to find and develop the next biomarkers or immunoreactive factors [35], which are related not only to antitumor effects but also treatment-refractory/resistant factors to prolong the survival of patients with gynecologic cancers.

## References

1. Liu J, Chen Z, Li Y, Zhao W, Wu J, Zhang Z. PD-1/PD-L1 checkpoint inhibitors in tumor immunotherapy. Front Pharmacol. 2021;12:731798. https://doi.org/10.3389/fphar.2021.731798.
2. Iwai Y, Hamanishi J, Chamoto K, Honjo T. Cancer immunotherapies targeting the PD-1 signaling pathway. J Biomed Sci. 2017;24(1):26. https://doi.org/10.1186/s12929-017-0329-9.
3. Vesely MD, Zhang T, Chen L. Resistance mechanisms to anti-PD cancer immunotherapy. Annu Rev Immunol. 2022;40:45–74. https://doi.org/10.1146/annurev-immunol-070621-030155.

4. Twomey JD, Zhang B. Cancer immunotherapy update: FDA-approved checkpoint inhibitors and companion diagnostics. AAPS J. 2021;23(2):39. https://doi.org/10.1208/s12248-021-00574-0.
5. Marabelle A, Le DT, Ascierto PA, Di Giacomo AM, De Jesus-Acosta A, Delord JP, et al. Efficacy of pembrolizumab in patients with noncolorectal high microsatellite instability/mismatch repair-deficient cancer: results from the phase II KEYNOTE-158 study. J Clin Oncol. 2020;38(1):1–10. https://doi.org/10.1200/JCO.19.02105.
6. Ishida Y, Agata Y, Shibahara K, Honjo T. Induced expression of PD-1, a novel member of the immunoglobulin gene superfamily, upon programmed cell death. EMBO J. 1992;11(11):3887–95. https://doi.org/10.1002/j.1460-2075.1992.tb05481.x.
7. Greenwald RJ, Freeman GJ, Sharpe AH. The B7 family revisited. Annu Rev Immunol. 2005;23:515–48. https://doi.org/10.1146/annurev.immunol.23.021704.115611.
8. Keir ME, Butte MJ, Freeman GJ, Sharpe AH. PD-1 and its ligands in tolerance and immunity. Annu Rev Immunol. 2008;26:677–704. https://doi.org/10.1146/annurev.immunol.26.021607.090331.
9. Okazaki T, Honjo T. PD-1 and PD-1 ligands: from discovery to clinical application. Int Immunol. 2007;19(7):813–24. https://doi.org/10.1093/intimm/dxm057. Epub 2007 Jul 2
10. Chung HC, Ros W, Delord JP, Perets R, Italiano A, Shapira-Frommer R, et al. Efficacy and safety of pembrolizumab in previously treated advanced cervical cancer: results from the phase II KEYNOTE-158 study. J Clin Oncol. 2019;37(17):1470–8. https://doi.org/10.1200/JCO.18.01265.
11. Tewari KS, Monk BJ, Vergote I, Miller A, de Melo AC, Kim HS, et al. EMPOWER-Cervical 1/GOG-3016/ENGOT-cx9: Interim analysis of phase III trial of cemiplimab vs. investigator's choice chemotherapy in recurrent/metastatic cervical carcinoma. Abstract presented at: ESMO Virtual Plenary, abstract VP4–2021; 2021 May 12; XXXXX, XX.
12. Goodman AM, Piccioni D, Kato S, Boichard A, Wang HY, Frampton G, et al. Prevalence of PDL1 amplification and preliminary response to immune checkpoint blockade in solid tumors. JAMA Oncol. 2018;4(9):1237–44. https://doi.org/10.1001/jamaoncol.2018.1701.
13. Colombo N, Dubot C, Lorusso D, Caceres MV, Hasegawa K, Shapira-Frommer R, et al. Pembrolizumab for persistent, recurrent, or metastatic cervical cancer. N Engl J Med. 2021;385(20):1856–67. https://doi.org/10.1056/NEJMoa2112435.
14. O'Malley DM, Randall LM, Jackson CG, Coleman RL, Hays JL, Moore KN, et al. RaPiDS (GOG-3028): randomized phase II study of balstilimab alone or in combination with zalifrelimab in cervical cancer. Future Oncol. 2021;17(26):3433–43. https://doi.org/10.2217/fon-2021-0529.
15. Taha T, Reiss A, Amit A, Perets R. Checkpoint inhibitors in gynecological malignancies: are we there yet? BioDrugs. 2020;34(6):749–62. https://doi.org/10.1007/s40259-020-00450-x.
16. Gubin MM, Zhang X, Schuster H, Caron E, Ward JP, Noguchi T, et al. Checkpoint blockade cancer immunotherapy targets tumour-specific mutant antigens. Nature. 2014;515(7528):577–81. https://doi.org/10.1038/nature13988.
17. Merelli B, Massi D, Cattaneo L, Mandalà M. Targeting the PD1/PD-L1 axis in melanoma: biological rationale, clinical challenges and opportunities. Crit Rev Oncol Hematol. 2014;89(1):140–65. https://doi.org/10.1016/j.critrevonc.2013.08.002.
18. Le DT, Durham JN, Smith KN, Wang H, Bartlett BR, Aulakh LK, et al. Mismatch repair deficiency predicts response of solid tumors to PD-1 blockade. Science. 2017;357(6349):409–13. https://doi.org/10.1126/science.aan6733.
19. Oaknin A, Tinker AV, Gilbert L, Samouëlian V, Mathews C, Brown J, et al. Clinical activity and safety of the anti-programmed death 1 monoclonal antibody dostarlimab for patients with recurrent or advanced mismatch repair-deficient endometrial cancer: a nonrandomized phase 1 clinical trial. JAMA Oncol. 2020;6(11):1766–72. https://doi.org/10.1001/jamaoncol.2020.4515.
20. Konstantinopoulos PA, Luo W, Liu JF, Gulhan DC, Krasner C, Ishizuka JJ, et al. Phase II study of avelumab in patients with mismatch repair deficient and mismatch repair proficient recurrent/persistent endometrial cancer. J Clin Oncol. 2019;37(30):2786–94. https://doi.org/10.1200/JCO.19.01021.
21. Antill YC, Kok PS, Robledo K, Barnes E, Friedlander M, Baron-Hay SE, et al. Activity of durvalumab in advanced endometrial cancer (AEC) according to mismatch repair (MMR) status:

the phase II PHAEDRA trial (ANZGOG1601). Abstract presented at: 2019 ASCO Annual Meeting, abstract 5501; 2019 May 31-Jun 4; Chicago, IL. https://ascopubs.org/doi/10.1200/JCO.2019.37.15_suppl.5501.
22. Makker V, Colombo N, Casado Herráez A et al. A multicenter, open-label, randomized, phase III study to compare the efficacy and safety of lenvatinib in combination with pembrolizumab versus treatment of physician's choice in patients with advanced endometrial cancer. Abstract presented at: Society of Gynecologic Oncology 52nd Annual Meeting on Women's Cancer, abstract 11512; 2021 18–21; Phoenix, AZ.
23. Lawrence MS, Stojanov P, Polak P, Kryukov GV, Cibulskis K, Sivachenko A, et al. Mutational heterogeneity in cancer and the search for new cancer-associated genes. Nature. 2013;499(7457):214–8. https://doi.org/10.1038/nature12213.
24. Marabelle A, Fakih M, Lopez J, Shah M, Shapira-Frommer R, Nakagawa K, Chung HC, Kindler HL, Lopez-Martin JA, Miller WH Jr, Italiano A, Kao S, Piha-Paul SA, Delord JP, McWilliams RR, Fabrizio DA, Aurora-Garg D, Xu L, Jin F, Norwood K, Bang YJ. Association of tumour mutational burden with outcomes in patients with advanced solid tumours treated with pembrolizumab: prospective biomarker analysis of the multicohort, open-label, phase 2 KEYNOTE-158 study. Lancet Oncol. 2020;21(10):1353–65. https://doi.org/10.1016/S1470-2045(20)30445-9. Epub 2020 Sep 10. PMID: 32919526. https://www.thelancet.com/journals/lanonc/article/PIIS1470-2045(20)30445-9/fulltext.
25. Association of tumour mutational burden with outcomes in patients with advanced solid tumours treated with pembrolizumab: prospective biomarker analysis of the multicohort, open-label, phase 2 KEYNOTE-158 study. Lancet Oncol. 2020;21(10):1353–65. https://doi.org/10.1016/S1470-2045(20)30445-9.
26. Rousseau B, Foote MB, Maron SB, Diplas BH, Lu S, Argilés G, et al. The spectrum of benefit from checkpoint blockade in hypermutated tumors. N Engl J Med. 2021;384(12):1168–70. https://doi.org/10.1056/NEJMc2031965.
27. McGrail DJ, Pilié PG, Rashid NU, Voorwerk L, Slagter M, Kok M, et al. High tumor mutation burden fails to predict immune checkpoint blockade response across all cancer types. Ann Oncol. 2021 May;32(5):661–72. https://doi.org/10.1016/j.annonc.2021.02.006.
28. Matulonis UA, Shapira-Frommer R, Santin AD, Lisyanskaya AS, Pignata S, Vergote I, et al. Antitumor activity and safety of pembrolizumab in patients with advanced recurrent ovarian cancer: results from the phase II KEYNOTE-100 study. Ann Oncol. 2019;30(7):1080–7. https://doi.org/10.1093/annonc/mdz135.
29. Myriad Genetics [Internet]. Salt Lake City: c2022. My Choice® CDx Myriad HRD Companion Diagnostic Test; [cited YYYY MMM DD]. Available from: https://myriad.com/oncology/mychoicecdx/.
30. Lee EK, Konstantinopoulos PA. Combined PARP and immune checkpoint inhibition in ovarian cancer. Trends Cancer. 2019;5(9):524–8. https://doi.org/10.1016/j.trecan.2019.06.004.
31. Drew Y, Kaufman B, Banerjee S, Lortholary A, Hong SH, Park YH, et al. Phase II study of olaparib + durvalumab (MEDIOLA): updated results in germline BRCA-mutated platinum-sensitive relapsed (PSR) ovarian cancer (OC). Ann Oncol. 2019;30(Suppl 5):v475–532.
32. Drew Y, Penson RT, O'Malley DM, Kim J, Zimmerman S, Roxburgh P, et al. 814MO - phase II study of olaparib (O) plus durvalumab (D) and bevacizumab (B) (MEDIOLA): initial results in patients (pts) with non-germline BRCA-mutated (non-gBRCAm) platinum sensitive relapsed (PSR) ovarian cancer (OC). Ann Oncol. 2020;31(Suppl 4):S551–89.
33. Wilson BG, Roberts CW. SWI/SNF nucleosome remodellers and cancer. Nat Rev Cancer. 2011;11(7):481–92. https://doi.org/10.1038/nrc3068.
34. Jelinic P, Ricca J, Van Oudenhove E, Olvera N, Merghoub T, Levine DA, Zamarin DJ. Immune-active microenvironment in small cell carcinoma of the ovary, hypercalcemic type: rationale for immune checkpoint blockade. Natl Cancer Inst. 2018;110(7):787–90. https://doi.org/10.1093/jnci/djx277.
35. Mills AM, Bullock TN, Ring KL. Targeting immune checkpoints in gynecologic cancer: updates & perspectives for pathologists. Mod Pathol. 2022;35(2):142–51. https://doi.org/10.1038/s41379-021-00882-y.

# Chapter 7
# Risk Assessment and Prevention Strategies for Hereditary Gynecological Cancers

**Sayaka Ueno and Akira Hirasawa**

**Abstract** A variety of hereditary cancer syndromes contribute to the development of gynecological cancers. These syndromes are caused due to germline pathogenic variants (GPVs) in tumor supressor genes or DNA repair genes. With the increasing use of genomic sequencing in clinical practice, the number of individuals diagnosed with GPVs in genes associated with hereditary cancer syndromes is increasing. Hereditary cancer syndromes differ in the types of cancer susceptible to develop, the risk of developing certain cancer, cancer treatment strategies, and possible cancer preventive strategies, depending on the gene responsible for the syndrome. Thus, physicians involved in the management of gynecological cancers perform accurate genetic risk assessments based on accurate knowledge about each syndrome and provide proper medical intervention to prevent developing cancer or to detect cancers in their early stage. Genetic risk assessments also helps in the selection of appropriate fertility preservation methods and treatment strategies for hormonal imbalances in women. Knowledge about significance and accuracy of various genetic tests may be helpful in interpreting the results of the test and in determining the appropriate medical interventions. Here, we reviewed mechanisms of cancer development and clinical features of hereditary gynecological cancers, as well as genetic risk assessment and cancer prevention strategies for those syndromes.

**Keywords** Hereditary gynecological cancers · Tumor suppressor genes · Loss of heterozygosity · Autosomal dominant inheritance · Genetic risk assessment · Genetic testing · Surveillance for cancer · Risk-reducing surgery

---

S. Ueno (✉) · A. Hirasawa
Department of Clinical Genomic Medicine, Graduate School of Medicine, Dentistry and Pharmaceutical Sciences, Okayama University, Okayama, Japan
e-mail: sayaka-u@okayama-u.ac.jp; hir-aki45@okayama-u.ac.jp



M. Mandai (ed.), *Personalization in Gynecologic Oncology*, Comprehensive Gynecology and Obstetrics, https://doi.org/10.1007/978-981-19-4711-7_7

## 7.1 Introduction

All cancers develop as a result of mutations in certain genes, such as those involved in the regulation of cell growth and/or DNA repair [1, 2]. Mutations can be classified into two types: germline mutations (recently described as germline variants) and somatic mutations. Germline variants can be passed on to the next generation and may be shared among relatives. Variants associated with certain diseases are defined as germline pathogenic variants. Some germline variants are the causes of hereditary cancer syndrome, which is defined as "a type of inherited disorder in which there is a higher-than-normal risk of certain types of cancer" according to the National Cancer Institute. Most hereditary cancer syndromes exhibit autosomal dominant inheritance, and the responsible genes are mostly tumor suppressor genes. By contrast, somatic mutations are acquired in somatic cells during their lifespan and are restricted to the individual in whom they occur.

*RB1* is the first human tumor suppressor gene to be described; it plays an integral role in the development of retinoblastoma. In 1993, the 180-kb genomic region encoding the RB1 transcript was sequenced; at the time, this was the longest stretch of human DNA sequence [3]. In the early 1990s, a number of tumor suppressor genes responsible for hereditary gynecological cancers were identified including *BRCA1*, *BRCA2*, *MLH1*, *MSH2*, *MSH6*, and *PMS2* [4–9]. *BRCA1/2* are most common causes of hereditary breast and ovarian cancers; *MLH1*, *MSH2*, *MSH6*, and *PMS2*, which are generally referred to as mismatch repair (MMR) genes, are responsible for Lynch syndrome. To date, more than 50 hereditary cancer syndromes have been described, and the responsible genes have been cloned.

Germline pathogenic/likely pathogenic variants (GPVs) were found in 8% of 10,389 adult cancer patients across 33 cancer types in the TCGA cohort [10]. The frequency of GPVs varied greatly among cancer types. In gynecological cancer, the prevalence rates of GPVs were 19.9% in ovarian serous cystadenocarcinoma, and 6.8% in uterine endometrial cancer (EC), and 6.6% in cervical cancer. The highest rate was observed in pheochromocytoma and paraganglioma (22.9%) followed by ovarian serous cystadenocarcinoma. Although not all of the GPVs identified were associated with the development of cancer that each individual was currently suffering from, the associations between *BRCA1/2* GPVs and ovarian cancer, *MSH6* and *PTEN* GPVs and EC were identified in this study.

This chapter summarizes the molecular mechanisms, clinical features, genetic risk assessment, and prevention strategies for hereditary gynecological cancers presented in Table 7.1.

**Table 7.1** Molecular and clinical features of hereditary gynecological cancers

| Syndrome | Responsible genes | Related gynecological cancers | Common histological subtypes | Other nongynecological tumors |
|---|---|---|---|---|
| *BRCA*-related breast/ovarian cancer syndrome | *BRCA1*, *BRCA2* | Ovarian cancer | Serous non-mutinous | Breast cancer, prostate cancer, pancreatic cancer |
| Lynch syndrome | *MLH1*, *MSH2*, *MSH6*, *PMS2*, *EPCAM* | Endometrial cancer | Endometrioid non-serous | Colorectal cancer, gastric cancer, small bowel cancer, urothelial cancer, pancreatic cancer |
| | | Ovarian cancer | Endometrioid | |
| PTEN hamartoma tumor syndrome (Cowden syndrome) | *PTEN* | Endometrial cancer | | Breast cancer, pancreatic cancer, colorectal cancer, gastric cancer, small bowel cancer, thyroid cancer |
| Peutz-Jeghers syndrome | *STK11* | Non-epithelial ovarian tumor | Sex cord tumor with annular tubules | Breast cancer, pancreatic cancer, colorectal cancer, gastric cancer, small bowel cancer |
| | | Cervical cancer | Gastric type mucinous carcinoma, LEGH | |
| DICER1 syndrome | *DICER1* | Non-epithelial ovarian tumor | Sertoli-Leydig cell tumor | Pleuropulmonary blastoma, pulmonary cysts, thyroid gland neoplasia, cystic nephroma |
| | | Cervical tumor | Embryonal rhabdomyosarcoma of the cervix | |
| Rhabdoid tumor predisposition syndrome | *SMARCA4* | Non-epithelial ovarian tumor | Hypercalcemic type of small cell carcinomas | Rhabdoid tumors of central nervous system, renal rhabdoid tumors |
| Other cancer-susceptible genes | *RAD51C*, *RAD51D* | Ovarian cancer | | Breast cancer |
| | *BRIP1* | | | Unknown |
| | *ATM* | | | Breast cancer, pancreatic cancer |
| | *PALB2* | | | Breast cancer, pancreatic cancer |

## 7.2 Biological Impacts of the Germline Variants in Hereditary Cancer

Cancer driver genes are classified as oncogenes or tumor suppressor genes, depending on whether their activation or inactivation contributes to cancer development. Although a single mutation in an oncogene can be sufficient for tumorigenesis, inactivation of both alleles of a tumor suppressor gene is often required.

In 1971, Alfred Knudson proposed the “two-mutation hypothesis” (now known as the two-hit theory), which states that in familial retinoblastoma cases, individuals

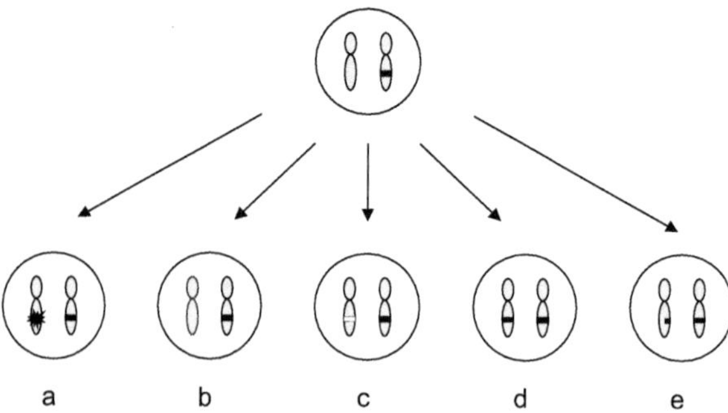


**Fig. 7.1** Various events account for the second hit in a cell with a pathogenic germline variant. (**a**) De novo mutation of the wild-type allele. (**b–d**) Three mechanisms of LOH: chromosomal loss, gene deletion, and somatic recombination (copy neutral LOH). Copy neutral LOH is a special case of LOH in which the wild-type allele is replaced with a mutant allele. (**e**) Promoter methylation of the wild-type allele. *LOH* loss of heterozygosity

possess one mutant RB allele due to an inherited or de novo germline mutation in the RB gene (first hit), and when a retina cell acquires a somatic mutation in the remaining wild-type allele (second hit), the cell will be transformed into a retinoblastoma cell [11]. The second hit described by Knudson could be accounted for via alternative molecular events, such as deletion of the wild-type allele, which is referred to as loss of heterozygosity (LOH), or DNA methylation changes in the wild-type allele (Fig. 7.1).

Although the patterns of somatic second-hit events differ depending on the tissue and genes, LOH is thought to be the most common second-hit event. LOH for the wild-type allele was reported in 92–100% and 70–76% of patients with germline *BRCA1* and *BRCA2* truncating variants in ovarian cancer [12, 13]. LOH events occurred more rarely in patients with germline missense variants of *BRCA1* and *BRCA2* than those with truncating variants, with a rate of 11% [13]. Cooperation between germline variants and somatically acquired alterations within not only the same gene but also different genes has been recently described in several tumor localizations [13]. In MMR gene-related cancer, LOH occurred in almost half of the patients with GPVs in MMR genes [14, 15]. Somatic single nucleotide variants were reported as the second most common mechanism of two-hit inactivation of MMR genes [14]. Another second-hit event, promoter methylation in *MLH1*, has been reported in colorectal cancer and ECs with *MLH1* GPVs [15, 16].

Although the two-hit theory is a clear model for explaining the contribution of tumor suppressor genes in tumorigenesis, even partial inactivation of tumor suppressor genes can also critically contribute to tumorigenesis [17]. In some tumor suppressor genes, a single copy of the wild-type allele is not enough to provide sufficient gene function, and thus called haploinsufficiency. Tumors in patients with Li-Fraumeni syndrome, which is caused by *TP53* GPVs, do not always exhibit loss of the wild-type *TP53* allele, suggesting that haploinsufficiency of *TP53* may be sufficient for tumor initiation [18]. *BRCA1/2* also show haploinsufficiency. Microscopically normal tissues in carriers of *BRCA1/2* GPVs have altered mRNA profiles compared with *BRCA* wild-type cells, suggesting an impact of one-hit events on tumorigenesis [19]. In addition, single-copy mutation of a tumor suppressor gene sometimes interferes with the function of the wild-type gene product, which is described as a dominant negative mutation. Certain missense variants in *ATM* have been reported to act in a dominant-negative manner to increase breast cancer risk, relative to truncating mutations [20–23].

## 7.3 Hereditary Gynecological Cancers

Gynecological cancers often overlap with hereditary cancer syndromes, therefore, gynecologists need to have a proper insight into hereditary cancer syndromes. The prevalence of GPVs in gynecological cancers and breast cancer is shown in Fig. 7.2. The frequency of GPVs in breast cancer patients was 9.9% [10]. About 10–20% of epithelial ovarian cancer patients are estimated to have GPVs in ovarian cancer susceptibility genes [24–26]. Some genes are associated with the development of non-epithelial ovarian cancer. About 5–10% of EC patients are estimated to have GPVs in EC-related genes [27–29]. Cervical cancer is in most cases caused by the human papillomavirus, and is thus very unlikely to be hereditary. To date, two types

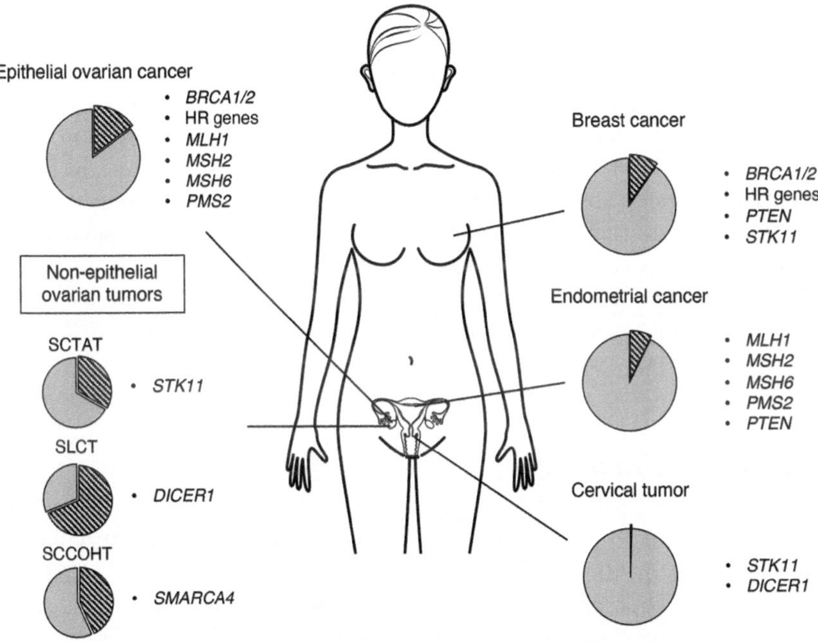


**Fig. 7.2** Prevalence of GPVs in breast, ovarian, endometrial and cervical cancers/tumors. The shaded area in the pie chart represents the probability of detecting GPVs in cancer susceptibility genes. Genes in which GPVs are commonly detected are listed on the right side of the pie chart. *GPV* germline pathogenic/likely pathogenic variants, *HR genes* genes involved in homologous recombination repair pathway; *ATM*, *BRIP1*, *PALB2*, *RAD51C*, *RAD51D*, etc.

of cervical cancer have been reported to be associated with hereditary tumors. This section outlines the typical gynecological hereditary cancers shown in Table 7.1.

### 7.3.1 BRCA-*Related Breast/Ovarian Cancer Syndrome (Hereditary Breast and Ovarian Cancer: HBOC)*

GPVs in *BRCA1/2* are associated with susceptibility to breast, ovarian, prostate, and pancreatic cancers. *BRCA1* and *BRCA2* are located on chromosome 17q21 and 13q12, respectively, and both genes encode proteins involved in DNA repair damage via the homologous recombination repair pathway and serve as tumor suppressors. The cumulative risks of developing breast and ovarian cancers by the age of 80 years are 72% and 44% for women with GPVs in *BRCA1*, 69% and 17% for those with GPVs in *BRCA2*, respectively [30].

GPVs in *BRCA1/2* are responsible for at least 10% of epithelial ovarian cancers [24, 31, 32]. Ovarian cancer in the context of *BRCA1/2* GPVs is characterized by a high proportion of serous carcinoma, advanced disease stage, and younger disease onset [24, 31–34].

It remains unknown whether *BRCA1/2* GPVs are associated with an increased risk of EC or not. A precious prospective cohort study showed a slightly increased risk of EC in a median follow-up of 5.7 years, with a standardized incidence ratio (SIR) of 1.91 (95% confidence interval [CI]: 1.06–3.19) for *BRCA1* carriers and 1.75 (95% CI: 0.55–4.23) for *BRCA2* carriers, which was not statistically significant [35]. In this study, tamoxifen use was identified as the most relevant risk factor for EC. Tamoxifen use significantly increased the SIR in *BRCA1* carriers from 1.91 to 4.43 (95% CI: 1.94–8.76), whereas in *BRCA2* carriers the association was not statistically significant (SIR = 2.29, 95% CI: 0.38–7.59). In another study including 1083 *BRCA1/2* carriers who underwent risk-reducing salpingo-oophorectomy (RRSO) without hysterectomy, the risk of developing EC did not increase within a median follow-up of 5.1 years [36]. However, of the eight incident uterine cancers observed, five were serous/serous-like and four of the five occurred in *BRCA1* carriers, indicating increased risk for serous/serous-like EC in *BRCA1* carriers.

### 7.3.2 *Lynch Syndrome*

Lynch syndrome (LS) is a hereditary cancer syndrome caused by GPVs in DNA mismatch repair (MMR) genes such as *MLH1*, *MSH2*, *MSH6*, and *PMS2* [37]. Additionally, deletion of the last exon of *EPCAM*, which is located upstream of *MSH2*, also causes LS through hypermethylation of the *MSH2* promoter and subsequent MSH2 silencing [38].

Individuals with LS are at a heightened risk of developing several types of cancers, which vary based on the affected MMR genes and age. An international, multicenter prospective observational study including 6350 participants with GPVs in

MMR genes showed that the cumulative risks of developing ECs by the age of 75 years were 37.0% for *MLH1*, 48.9% for *MSH2*, 41.1% for *MSH6*, and 12.8% for *PMS2* carriers [39]. For ovarian cancer, the cumulative risks were 11.0% for *MLH1*, 17.4% for *MSH2*, 10.8% for *MSH6*, and 3.0% for *PMS2* carriers.

Gynecological cancers in the context of LS are mainly EC and characterized by a younger disease onset [40–42]. The prevalence rates of LS have been reported to be 5.8–7.2% in EC patients [28, 29], and 0.4–3% in epithelial ovarian cancer patients [24, 43, 44]. Synchronous endometrial and ovarian cancers were reported in 21.6% of LS-associated EC patients and also in LS-associated ovarian cancer patients [40, 45]. In 81.4% of individuals with LS, EC was first cancer in that individuals. The lower uterine segment was involved in 25% of LS-associated EC patients [40].

### *7.3.3 PTEN Hamartoma Tumor Syndrome (Cowden Syndrome)*

PTEN hamartoma tumor syndrome is a multiple hamartoma syndrome frequently associated with GPVs in *PTEN* [46]. *PTEN*, located on chromosome 10q23, encodes a phosphatase involved in cell signaling pathways that affect cell proliferation and survival.

Hamartomas are benign tumors that result from overgrowth of normal tissues. Multiple hamartomas occurring in various organs are a common manifestation of this syndrome. Individuals with this syndrome often exhibit other characteristic features, such as macrocephaly and multiple mucocutaneous lesions, therefore, most patients would be clinically diagnosed.

This syndrome is also associated with an increased risk of developing several types of cancer, including breast, endometrial, thyroid, and colorectal cancer. Among all, breast cancer is the most common type of cancer in patients with this syndrome, with a lifetime risk of up to 85% [47]. The lifetime risk of developing EC is estimated to be 28%, with the risk beginning to increase at the age of 25 years and rising to 30% by the age of 60 years [28, 47].

### *7.3.4 Peutz-Jeghers Syndrome*

Peutz-Jeghers syndrome (PJS) is characterized by multiple hamartoma polyps in the gastrointestinal tract, pigmentation of the skin mucosa as well as increased susceptibility to cancer in the gastrointestinal tract, uterine cervix, testes, ovary, and breast [48, 49]. Most of the PJS cases are due to GPVs in the *STK11* (*LKB1*) gene [50, 51]. *STK11*, located on chromosome 19p13, encodes a serine-threonine kinase involved in cell polarity, metabolism, and growth.

Gynecological tumors associated with PJS are sex cord tumor with annular tubules (SCTAT) of ovary and cervical gastric type mucinous carcinoma of the endocervix (G-ECA). The lifetime risks of developing SCTAT and G-ECA was reported to be 21% and 10%, respectively, with the average ages at diagnosis of 28 years for SCTAT and 34–40 years for G-ECA [49, 52]. Among all patients with ovarian SCTAT, approximately one-third have PJS [53]. PJS-related G-ECAs are extremely well-differentiated forms of G-ECA known as adenoma malignum or minimal deviation adenocarcinoma (MDA). Among patients with MDA, 11–17% have PJS [54, 55]. Although lobular endocervical glandular hyperplasia (LEGH) is a basically benign gastric type mucinous lesion of cervix, LEGH with atypia could be a precursor of MDA [56]. The first case of LEGH in a patient with a *STK11* GPV who was diagnosed PJS was reported in 2012 [57]. Since then, a few case reports have shown that LEGH can be associated with PJS [58–60].

### 7.3.5 DICER1 Syndrome

DICER1 syndrome is characterized by pediatric pleuropulmonary blastoma, nodular hyperplasia of the thyroid, cystic nephroma, Sertoli-Leydig cell tumors of the ovary (SLCT), and other rare types of tumors [61, 62]. This syndrome is caused by GPVs in *DICER1*, located on chromosome 14q32, which encodes an RNase III endonuclease involved in posttranscriptional gene expression by modulating microRNAs [63, 64]. In most cases, biallelic variants in *DICER1* have been detected in tumors: usually a loss-of-function GPV in one allele and a tumor-specific somatic hotspot variant in the second allele [65]. Monoallelic loss of *DICER1* can promote tumorigenesis, indicating its haplo-insufficient function as a tumor suppressor gene [66].

The lifetime risk of developing SLCTs was estimated to be 21.2% with the average age at diagnosis of 16.9 years [67, 68]. In SLCT patients, *DICER1* GPVs were identified in 18 of 26 patients (69%) [69].

Embryonal rhabdomyosarcoma of the cervix (cERMs) is a rare type of tumor that occurs in older children, adolescents and young adults with a median age of 13–14 years [70]. The association between cERMs and SLCT was later reported in a cohort of 14 patients [71]. Although the lifetime risk of developing cERMs in *DICER1* carriers has not been reported, most of the cERMs (18 of 19 patients, 95%) were reported to have *DICER1* mutations, 50% of which were of germline origin (6 of 12 patients tested) [72].

### 7.3.6 Rhabdoid Tumor Predisposition Syndrome

*SMARCA4*, located on chromosome 19p13, is a chromatin remodeling gene and encodes BRG1. Recently, biallelic inactivation of *SMARCA4* and the consequent complete loss of BRG1 protein have been identified as molecular event defining small cell carcinoma of the ovary, hypercalcemic type (SCCOHT) [73–75].

*SMARCA4* GPVs were identified in 43% of SCCOHT patients (26/60), with significantly younger age at diagnosis than those without GPVs [76]. *SMARCA4* carriers also develop rhabdoid tumors involving the central nervous system or kidneys [77]. Since the incidence of GPVs is high, the International SCCOHT Consortium recommends referral of all patients with SCCOHT to a clinical genetics service and offering genetic tests for *SMARCA4* GPVs [78].

### 7.3.7 *Other Cancer-Susceptible Genes*

Recently, several genes that are involved in the development of hereditary ovarian cancers have been identified. Compared with *BRCA1/2* and MMR genes, the penetrance of these genes is lower, but not negligible. Among these genes, *ATM*, *BRIP1*, *PALB2*, *RAD51C*, and *RAD51D* are involved in the homologous recombination repair pathway as well as *BRCA1/2*.

*ATM* GPVs were found in 0.64–0.87% of ovarian cancer patients, which was significantly greater than the 0.1% frequency in healthy controls [79]. *ATM* GPVs were estimated to slightly increase the risk of developing ovarian cancer [80].

*BRIP1* GPVs were found in about 1% of ovarian cancer patients [24, 25]. A previous large case control study showed that *BRIP1* is associated with an increased risk of developing ovarian cancer, especially high-grade serous ovarian cancer, with a relative risk of 14.09 (95% CI, 4.04–45.02, $p < 0.001$). In *BRIP1* carriers, the cumulative lifetime risk of developing ovarian cancer by the age of 80 years was estimated to be 5.8% [81].

*PALB2* GPVs were found in about 0.38–0.62% of ovarian cancer patients [24, 25]. Whether *PALB2* GPVs increase the risk of developing ovarian cancer remains unknown. Although two previous studies demonstrated an association, three other studies did not show a statistically significant association between *PALB2* GPVs and increased ovarian cancer risk [24, 81–84].

*RAD51C* and *RAD51D* GPVs were found in about 0.5% of ovarian cancer patients respectively [24, 25]. Previous case control studies identified an association between *RAD51C* and *RAD51D* GPVs and increased ovarian cancer risk, with odds ratios of 3.4–5.2 and 4.78–12.0, respectively [24, 83, 85].

## 7.4 Genetic Risk Assessment

The typical clinical features of hereditary cancers are as follows: (1) younger age of onset, (2) accumulation of certain types of cancers in the family members, (3) presence of multiple types of cancer in one person, and (4) occurrence of cancer in both paired organs. The purpose of genetic risk assessment is to identify the individuals who may be at risk of hereditary cancer syndromes and may benefit from genetic testing, additional screening, or preventive medical interventions. In many cases, gynecologists will play an important role in the identification and referral of women at risk for these conditions. In this section, we will summarize the clues for evaluating the personal risk of hereditary cancer syndromes.

### 7.4.1 Personal and Family History of Cancer

Collecting a detailed personal and family history is the first step in genetic risk assessment. Accurate genetic risk assessment requires, at a minimum, family history of first- and second-, and hopefully third-degree relatives of both maternal and paternal sides. Personal and family history will change over time; therefore, clinicians are required to update the data. History of cancer should be collected, including age at diagnosis, subtype, pathology, and laterality of the disease. Surgical history, such as salpingo-oophorectomy for benign ovarian tumors or total hysterectomy for uterine myomas, is an important information since these may serve as risk-reducing surgeries for ovarian or endometrial cancers. Hormonal therapy history, the use of oral contraceptive, carcinogen exposure history, and ethnic background can also influence the results of genetic risk assessment.

To identify candidates for genetic services, clinicians can use published categorical guidelines available through professional organizations [86–90]. In addition, some models are provided to predict the probability that an individual has GPVs in *BRCA1/2* or any of the MMR genes. These include the BRCAPRO and BOADICEA models in *BRCA1/2* and the PREMM5, MMRpredict, and MMRpro for MMR genes [91–95]. Because each model is developed based on a study of a certain population, the use of these models is appropriate only when the patient's characteristics and family history are similar to those of the study population. Ethnicity, the histology of cancer, and laterality of cancer can influence the accuracy of the models [96–100]. In addition, BRCAPRO was insufficient to predict *BRCA1/2* GPVs in ovarian cancer patients [101].

### 7.4.2 Characteristic Physical Findings Other than Cancer

Some hereditary cancer syndromes are accompanied by distinctive clinical findings other than the development of certain cancers. Detection of trichilemmomas or oral mucosal papillomatosis on dermatologic examination, macrocephaly on measurement of head circumference, and multinodular goiter on thyroid palpation can be helpful in the diagnosis of PTEN hamartoma tumor syndrome (Cowden syndrome). In addition, hamartomas or esophageal glycogenic acanthoses can be detected incidentally during gastrointestinal endoscopy.

Hyperpigmentation of the mouth, lips, nose, eyes, genitalia, or fingers on inspection, or hamartomatous polyps of the gastrointestinal tract on endoscopy can be helpful in the diagnosis of PJS.

### 7.4.3 Result of Prior Genetic Tests in Family

The results of prior genetic tests of other family members would be helpful for the assessment. If a GPV has already been identified in other family members, searching only for the same location in the gene can be a reasonable and cost-effective diagnostic approach. However, more than one GPV may be present in a single family; thus, broader testing should be considered if multiple GPVs are suspected.

Pharmacogenetic tests, such as microsatellite instability (MSI) testing of tumor tissue, tumor testing for homologous recombination deficiency (HRD), or tumor clinical sequencing, could reveal the possibility of hereditary cancers. LS was identified in 16.3% of patients with MSI-high tumors [102]. BRCA1/2 play central roles in the homologous recombination pathway; thus, the HRD status indicates the possibility of *BRCA1/2* GPVs. GPVs of other genes involved in the homologous recombination pathway may cause HRD. Mutations found in clinical tumor sequencing could be of germline origin; therefore, offering opportunity to take the confirmation tests should be considered [103].

These results should be obtained from laboratories certificated for genetic testing. Recently, the genetic test results obtained through direct-to-consumer (DTC) services have been increasing. DTC genetic testing can be performed directly by an individual because DNA sampling from oral mucosa or hair is easily performed as it does not require for special equipment and is usually less expensive than clinical genetic testing. Given the limited testing methods and the higher rate of false-positive and false-negative results compared with clinical genetic testing, the results of DTC genetic testing should be re-evaluated by experts in genetics [104].

### *7.4.4 Clinical use of Multigene Panel Testing*

Historically, genetic testing for cancer patients has been conducted by first inferring the most likely hereditary cancer syndromes based on genetic risk assessment, and then testing for the single genes associated with these syndromes.

Genetic risk assessment plays an important role in the identification of individuals at risk of hereditary cancer syndrome, however, multiple factors may influence the accuracy of assessment. These factors include small family size, unknown family history, early deaths, and de novo pathogenic variants. In addition, with the rapid advances in sequencing technology, a number of genes with low to moderate cancer susceptibility have been identified. This variability in the penetrance of pathogenic variants may influence the risk assessment as well as the patterns of inheritance and mosaicism.

Moreover, several studies have reported that GPVs in cancer predisposition genes were identified not only in those who met the previous National Comprehensive Cancer Network (NCCN) testing criteria based on the genetic risk assessment but also in those who did not meet the criteria [105, 106]. Another retrospective analysis showed that only 18.9% of positive results in genetic test were consistent with the suspected syndromes and associated genes [107].

Now, next generation sequencing technology has enabled the simultaneous testing of a set of genes at low cost, that is, a multigene panel testing (MGPT). The introduction of MGPT should increase the number of individuals diagnosed with GPVs in hereditary cancer-associated genes that cannot be identified by conventional single gene tests. Indeed, in clinical settings, with growing evidence showing that certain genes other than *BRCA1/2* confer an increased risk of cancer predisposition, MGPT replaced the *BRCA1/2*-only tests in 2014 [108]. In 2020, the NCCN guidelines underwent a major paradigm shift by changing the description to consider MGPT first among genetic tests.

As mentioned above, MGPT is a useful and cost-effective tool for diagnosing hereditary cancer syndromes. However, for many of genes with low to moderate cancer susceptibility, only limited data are available on the degree of cancer risk, and no clear guidelines on risk management have been established. Therefore, medical intervention for individuals with GPVs in these genes should be considered based on the results of genetic risk assessment; genetic risk assessment remains important in management of hereditary cancer syndromes.

## 7.5 Cancer Prevention Strategies for Hereditary Cancer Syndromes

Individuals who are presumed to be at risk of hereditary cancer syndromes or who are concerned about these syndromes should be provided with the opportunity to receive genetic counseling prior to making any decisions regarding genetic testing. Genetic counseling has been defined as "the process of helping people understand and adapt to the medical, psychological, and familial implications of genetic contributions to disease" [109]. Through this process, individuals will be informed about the genes they may be tested, possible results and medical management associated with the results, and the implications of genetic testing for other family members. The benefits, risks, and limitations of genetic testing should also be discussed. This process facilitates informed decision-making and adaptation to the results of genetic testing.

Genetic testing is not always necessary for individuals who have already been diagnosed with certain hereditary cancer syndromes according to the clinical diagnostic criteria, as in most of such cases, the results of the test will not change medical management. Though, if a GPV was identified in the individual diagnosed with the disease, this information can also be used for genetic testing in other family members and can help predicting the inheritance manner. As such, identified genetic information can be information that can be of medical or psychological benefit to family members. The results of genetic testing should be carefully evaluated and disclosed to individuals along with the medical management options that could be offered to them. In this section, the recommended cancer risk management based on the genetic test results are summarized (Table 7.2).

**Table 7.2** Prevention strategies for hereditary gynecological tumors

| Gene | Gynecological organs | Screening[a] | Risk-reducing surgery | Other options |
|---|---|---|---|---|
| *BRCA1*, *BRCA2* | Ovary | Consider serum CA125 and TVUS | Recommend salpingo-oophorectomy | Oral contraceptives |
| *MLH1*, *MSH2*, *MSH6*, *PMS2*, *EPCAM* | Endometrial | Consider endometrial biopsy and TVUS | Consider hysterectomy | |
| | Ovary | Consider serum CA125 and TVUS | *MLH1*, *MSH2*, *EPCAM*: Consider salpingo-oophorectomy *MSH6*, *PMS2*: Insufficient evidence to recommend | |
| *PTEN* | Endometrial | Consider endometrial biopsy and TVUS | Discuss option of hysterectomy | |
| *STK11* | Cervix/Ovary | Annual pelvic exam with annual pelvic ultrasound and pap smear<br>Endometrial biopsy if abnormal bleeding | Consider hysterectomy | |
| *BRIP1* | Ovary | Consider serum CA125 and TVUS | Consider salpingo-oophorectomy | |
| *RAD51C*, *RAD51D* | Ovary | Consider serum CA125 and TVUS | Consider salpingo-oophorectomy | |
| *ATM* | Ovary | Consider serum CA125 and TVUS | Consider salpingo-oophorectomy based on family history | |
| *PALB2* | Ovary | Consider serum CA125 and TVUS | Consider salpingo-oophorectomy based on family history | |

*TVUS* transvaginal ultrasound
Table was created based on NCCN Guidelines Genetic/Familial High-Risk Assessment: Colorectal Version 1.2022, and Genetic/Familial High-Risk Assessment: Breast, Ovarian, and Pancreatic Version 2.2022
[a]Screening for ovarian cancer and endometrial cancer is of uncertain benefit

### *7.5.1* BRCA1/2

*BRCA1/2* GPV carriers have an extremely high risk of developing breast and ovarian cancers, as well as an increased risk for pancreatic and prostate cancers.

As *BRCA1/2* GPVs are associated with early-onset breast cancer, breast cancer screening should be initiated earlier than the standard recommendation [110]. For women with *BRCA1/2* GPVs, training in breast awareness starting at the age of 18 years, clinical breast examination every 6–12 months and annual breast MRI screening with contrast starting at the age of 25 years, and additional annual mammography with consideration of tomosynthesis beginning at the age of 30 years are recommended. In a prospective screening trial evaluating the performance of annual MRI and mammography in women with *BRCA1/2* GPVs, the sensitivity of MRI was significantly higher than that of mammography [111]. Furthermore, the majority or cancers detected by MRI screening are early-stage tumors. Another study reported that breast MRI had sensitivity rates of 79% for all cancers and 88.5% for invasive cancers, and a specificity rate of 86% [112]. Risk-reducing mastectomy (RRM) reduces the risk of developing breast cancer, although there is still no consensus on whether RRM reduces mortality. Therefore, the option of RRM should be carefully discussed during genetic counseling.

In contrast to breast cancer, RRSO is the current standard of care for ovarian cancer risk management in women with *BRCA1/2* GPVs [88, 113, 114]. In patients with *BRCA1/2* GPVs, the effectiveness of RRSO in reducing the risk of ovarian or fallopian cancer was reported to be 80–85%, with reduced mortality [115–117]. RRSO may provide an opportunity to detect clinically occult gynecologic cancers, especially serous tubal intraepithelial carcinoma (STIC), which is considered to be an early precursor lesion for serous ovarian cancers, in approximately 5–8% of patients [118, 119].

As described above, RRSO is an effective approach to reduce the risk of ovarian cancer in patients with *BRCA1/2* GPVs. However, before deciding to undergo RRSO, several topics should be discussed, such as the reproductive impact, residual risk of peritoneal cancer, and premature menopause. Even after RRSO, a 1–4.3% risk of developing peritoneal carcinoma remains, with the older age at RRSO and the presence of STIC in the RRSO specimen as the risk factors [120, 121]. Premenopausal women who undergo RRSO will experience acute climacteric symptoms of hormonal withdrawal.

Hormone replacement treatment (HRT) will not only attenuate these symptoms, but will also prevent the occurrence of osteoporosis and cognitive decline and help maintain cardiovascular health. HRT after RRSO for a short period has no reported effect on the breast cancer risk [122, 123]. Another study showed that short-term HRT use (mean duration: 4.3 years) did not increase breast cancer risk in female *BRCA1* GPV carriers without RRSO [124]. Although there have been no data about association between long-term use of HRT in *BRCA1/2* GPV carriers and breast cancer risk, in general population, the long-term use of HRT (median: 5.6 years) was associated with higher breast cancer incidence [125]. Therefore, information on the benefits and risks of HRT in individuals with *BRCA1/2* GPVs should be provided to them and the choice of whether to use HRT and for how long should be carefully discussed.

Salpingectomy with delayed oophorectomy could be another option for premenopausal women. Although several studies have shown the safety and feasibility of this procedure, more data are needed to determine its efficacy in reducing the risk of ovarian cancer [126, 127]. For those who have not elected RRSO, screening with transvaginal ultrasound and measurement of serum CA-125 levels may be considered in the clinical setting, although the clinical benefits remain uncertain.

The use of oral contraceptives (OCs) was reported to reduce the cumulative incidence of ovarian cancer from 1.2% to a maximum of 0.7% in general population; the incidence became lower the longer the OCs were used [128]. Three meta-analysis studies showed that the use of OCs reduces the risk of developing ovarian cancer by approximately 50% in *BRCA1/2* carriers [129–131].

Previous data showed conflicting data on the effect of OC use on breast cancer risk among *BRCA1/2* carriers [132–135]. Two meta-analyses showed no significant association between OC use and breast cancer risk in *BRCA1/2* carriers [129, 131]. Taken together, OC can be used to prevent ovarian cancer risk; however, physicians should be aware that the preventive effect is smaller than that of RRSO, and the appropriate duration of OC use remains uncertain.

Men with *BRCA1/2* GPVs have an increased risk of developing breast cancer, with the cumulative lifetime risks of 1.2% for those with *BRCA1* GPVs and 7–8% for those with *BRCA2* GPVs, compared with the cumulative lifetime risk of 0.1% in the general population [136–139]. For men with *BRCA1/2* GPVs, training in breast self-examination starting at age of 35 years is recommended, while starting annual mammography should be considered at age 50 or 10 years prior to the earliest known breast cancer in the family for those with gynecomastia.

Men with *BRCA1/2* GPVs also have an increased risk of developing prostate cancer [140–143]. Prostate cancer in male *BRCA1/2* carriers were often at an advanced or metastatic stage. Screening for prostate cancer using serum PSA starting at the age of 40 years should be recommended for those with *BRCA2* GPVs and should be considered for those with *BRCA1* GPVs [142].

If at least one first- or second-degree relative developed pancreatic cancer, pancreas cancer screening may be considered [144]. Pancreas cancer screening contributes to the earlier detection of pancreatic cancer and the improvement of resection rates, which may decrease the mortality rate [145, 146]. Screening may be performed using contrast-enhanced MRI/MRCP and/or endoscopic ultrasound starting at the age of 50 years or 10 years younger than the earliest pancreatic cancer diagnosis in the family [144].

### 7.5.2 MMR Genes (Lynch Syndrome)

Individuals with LS have an increased lifetime risk of developing several types of cancers, particularly colorectal and endometrial cancer. Although different genes carry different risks, the lack of large-scale cohort studies on the risks among specific variant carriers has resulted in the application of the same management at present.

Annual or semiannual colonoscopy starting at the age of 20–25 years or 2–5 years younger than the youngest diagnosis age in the family is recommended [147–152].

In women with LS, endometrial cancer is the second most common type of cancer, with a lifetime risk of up to approximately 50%; the risk varies by gene [39]. Due to the lack of sufficient evidence for specific routine screening, uniform guidelines for the surveillance of endometrial cancer in patients with LS are not currently available. However, in the clinical setting, endometrial biopsy in combination with transvaginal ultrasound is often performed with the expectation of improving the rate of endometrial cancer detection [153–155]. Women with LS are also at a higher risk of developing ovarian cancer. However, there has been no data supporting routine screening for ovarian cancer. Total hysterectomy and bilateral salpingo-oophorectomy can be performed as risk-reducing surgery [156].

There is no clear evidence to support the appropriate method for screening other types of cancer, including gastric, small bowel, urothelial, and pancreatic cancer. However, individuals with a familial history of each cancer may benefit from upper endoscopy, urinalysis, or imaging of the pancreas using MRI/MRCP or EUS. Recently, a PSA screening study in those with GPVs in MMR genes was conducted, demonstrating a higher prostate cancer incidence in *MSH2* and *MSH6* GPV carriers than in noncarrier controls and the usefulness of PSA screening in detecting prostate cancer [157].

### *7.5.3* PTEN *(PTEN Hamartoma Tumor Syndrome/ Cowden Syndrome)*

In PTEN hamartoma tumor syndrome, the cumulative lifetime risk for any types of cancer is estimated to be more than 80%, with a twofold greater cancer risk in women compared with that in men [158, 159]. The recommended screening strategy for breast cancer is similar to that for *BRCA1/2* GPV carriers. Although there has been no data regarding the efficacy of risk reduction surgery for breast cancer, RRM could be an option for women with this syndrome. For endometrial cancer, no study has reported the efficacy of screening; however, endometrial biopsy combined with transvaginal ultrasound could be considered. An annual thyroid ultrasound starting at the age of 7 years should be performed [160]. For risks of other cancers, colonoscopy, renal ultrasound, or upper endoscopy should be considered.

### *7.5.4* STK11 *(Peutz-Jeghers Syndrome)*

Individuals with this syndrome have increased risks of developing several types of cancers, including colorectal, breast, pancreatic, ovarian and gallbladder cancer. Surveillance for the multiple organs mentioned above is recommended, although there exist limited data regarding the efficacy of the screening modalities in this syndrome. For cervical and ovarian cancer, annual pelvic examination and pap smear should be considered. Pap smear alone reported to have limited diagnostic power for PJS-related cervical neoplasm, therefore, combination of MRI, Pap smears, and testing for gastric mucin may improve the accuracy of diagnosis [161].

### *7.5.5* BRIP1/RAD51C/RAD51D/ATM/PALB2

These genes are involved in the homologous recombination repair pathway as well as *BRCA1/2*, therefore, the risk prevention strategies for ovarian cancer should be similar to those for *BRCA1/2*.

Among them, *BRIP1*, *RAD51C*, and *RAD51D* are associated with a relatively higher risk of ovarian cancer, with estimated lifetime risk of over 10%. Therefore, RRSO should be considered in individuals with GPVs of these genes, although the optimal age for surgery remains unclear. Since the risk of ovarian cancer in *ATM* and *PALB2* GPV carriers is estimated to be relatively low, RRSO might be an option, depending on the family history.

## 7.6 Conclusions

Recent advances in DNA sequencing technology and development of molecularly targeted drugs have increased opportunity to identify GPVs in cancer-susceptible genes. Whole exome and genome sequencing, which will be used in clinical practice in near future, will further increase such opportunities. Genetic information will not change over lifetime, can predict the onset of disease, and may be shared with blood relatives. Hence, diagnosing an individual with hereditary cancer syndrome is equivalent to diagnosing an entire family with a hereditary cancer syndrome.

To know the genetic information will be the first step toward preventing cancer in families with hereditary cancer syndromes. The second step will be to understand the exact risk of developing susceptible cancers and preventive strategies for these conditions, and the third will be to share the genetic information with at-risk relatives. As gynecologists will be involved in each of these steps, it is essential to be familiar with gynecological hereditary cancers. Thus, gynecologists are encouraged to perform proper assessment of genetic risk, provide accurate information about the syndromes, and discuss with the patients how to share and effectively use the genetic information obtained for the health management of other family members. Last but not least, to collaborate with specialists in other departments is also important as multiple organs other than gynecological organs are involved in hereditary cancer syndrome.

**Conflict of interest** The authors declare no competing interests.

## References

1. Fearon ER, Vogelstein B. A genetic model for colorectal tumorigenesis. Cell. 1990;61(5):759–67. https://doi.org/10.1016/0092-8674(90)90186-i.
2. Vogelstein B, Kinzler KW. The multistep nature of cancer. Trends Genet. 1993;9(4):138–41. https://doi.org/10.1016/0168-9525(93)90209-z.
3. Toguchida J, McGee TL, Paterson JC, Eagle JR, Tucker S, Yandell DW, et al. Complete genomic sequence of the human retinoblastoma susceptibility gene. Genomics. 1993;17(3):535–43. https://doi.org/10.1006/geno.1993.1368.
4. Miki Y, Swensen J, Shattuck-Eidens D, Futreal PA, Harshman K, Tavtigian S, et al. A strong candidate for the breast and ovarian cancer susceptibility gene BRCA1. Science. 1994;266(5182):66–71. https://doi.org/10.1126/science.7545954.

5. Wooster R, Bignell G, Lancaster J, Swift S, Seal S, Mangion J, et al. Identification of the breast cancer susceptibility gene BRCA2. Nature. 1995;378(6559):789–92. https://doi.org/10.1038/378789a0.
6. Bronner CE, Baker SM, Morrison PT, Warren G, Smith LG, Lescoe MK, et al. Mutation in the DNA mismatch repair gene homologue hMLH1 is associated with hereditary nonpolyposis colon cancer. Nature. 1994;368(6468):258–61. https://doi.org/10.1038/368258a0.
7. Leach FS, Nicolaides NC, Papadopoulos N, Liu B, Jen J, Parsons R, et al. Mutations of a mutS homolog in hereditary nonpolyposis colorectal cancer. Cell. 1993;75(6):1215–25. https://doi.org/10.1016/0092-8674(93)90330-s.
8. Nicolaides NC, Papadopoulos N, Liu B, Wei YF, Carter KC, Ruben SM, et al. Mutations of two PMS homologues in hereditary nonpolyposis colon cancer. Nature. 1994;371(6492):75–80. https://doi.org/10.1038/371075a0.
9. Palombo F, Gallinari P, Iaccarino I, Lettieri T, Hughes M, D'Arrigo A, et al. GTBP, a 160-kilodalton protein essential for mismatch-binding activity in human cells. Science. 1995;268(5219):1912–4. https://doi.org/10.1126/science.7604265.
10. Huang KL, Mashl RJ, Wu Y, Ritter DI, Wang J, Oh C, et al. Pathogenic germline variants in 10,389 adult cancers. Cell. 2018;173(2):355–70.e14. https://doi.org/10.1016/j.cell.2018.03.039.
11. Knudson AG Jr. Mutation and cancer: statistical study of retinoblastoma. Proc Natl Acad Sci U S A. 1971;68(4):820–3. https://doi.org/10.1073/pnas.68.4.820.
12. Kanchi KL, Johnson KJ, Lu C, McLellan MD, Leiserson MD, Wendl MC, et al. Integrated analysis of germline and somatic variants in ovarian cancer. Nat Commun. 2014;5:3156. https://doi.org/10.1038/ncomms4156.
13. Lu C, Xie M, Wendl MC, Wang J, McLellan MD, Leiserson MD, et al. Patterns and functional implications of rare germline variants across 12 cancer types. Nat Commun. 2015;6:10086. https://doi.org/10.1038/ncomms10086.
14. Porkka N, Valo S, Nieminen TT, Olkinuora A, Mäki-Nevala S, Eldfors S, et al. Sequencing of Lynch syndrome tumors reveals the importance of epigenetic alterations. Oncotarget. 2017;8(64):108020–30. https://doi.org/10.18632/oncotarget.22445.
15. Ollikainen M, Hannelius U, Lindgren CM, Abdel-Rahman WM, Kere J, Peltomäki P. Mechanisms of inactivation of MLH1 in hereditary nonpolyposis colorectal carcinoma: a novel approach. Oncogene. 2007;26(31):4541–9. https://doi.org/10.1038/sj.onc.1210236.
16. Moreira L, Muñoz J, Cuatrecasas M, Quintanilla I, Leoz ML, Carballal S, et al. Prevalence of somatic mutl homolog 1 promoter hypermethylation in Lynch syndrome colorectal cancer. Cancer. 2015;121(9):1395–404. https://doi.org/10.1002/cncr.29190.
17. Berger AH, Knudson AG, Pandolfi PP. A continuum model for tumour suppression. Nature. 2011;476(7359):163–9. https://doi.org/10.1038/nature10275.
18. Varley JM, Evans DG, Birch JM. Li-Fraumeni syndrome—a molecular and clinical review. Br J Cancer. 1997;76(1):1–14. https://doi.org/10.1038/bjc.1997.328.
19. Bellacosa A, Godwin AK, Peri S, Devarajan K, Caretti E, Vanderveer L, et al. Altered gene expression in morphologically normal epithelial cells from heterozygous carriers of BRCA1 or BRCA2 mutations. Cancer Prev Res (Phila). 2010;3(1):48–61. https://doi.org/10.1158/1940-6207.Capr-09-0078.
20. Chenevix-Trench G, Spurdle AB, Gatei M, Kelly H, Marsh A, Chen X, et al. Dominant negative ATM mutations in breast cancer families. J Natl Cancer Inst. 2002;94(3):205–15. https://doi.org/10.1093/jnci/94.3.205.
21. Hall MJ, Bernhisel R, Hughes E, Larson K, Rosenthal ET, Singh NA, et al. Germline pathogenic variants in the ataxia telangiectasia mutated (ATM) gene are associated with high and moderate risks for multiple cancers. Cancer Prev Res (Phila). 2021;14(4):433–40. https://doi.org/10.1158/1940-6207.Capr-20-0448.
22. Southey MC, Goldgar DE, Winqvist R, Pylkäs K, Couch F, Tischkowitz M, et al. PALB2, CHEK2 and ATM rare variants and cancer risk: data from COGS. J Med Genet. 2016;53(12):800–11. https://doi.org/10.1136/jmedgenet-2016-103839.

23. Goldgar DE, Healey S, Dowty JG, Da Silva L, Chen X, Spurdle AB, et al. Rare variants in the ATM gene and risk of breast cancer. Breast Cancer Res. 2011;13(4):R73. https://doi.org/10.1186/bcr2919.
24. Norquist BM, Harrell MI, Brady MF, Walsh T, Lee MK, Gulsuner S, et al. Inherited mutations in women with ovarian carcinoma. JAMA Oncol. 2016;2(4):482–90. https://doi.org/10.1001/jamaoncol.2015.5495.
25. Carter NJ, Marshall ML, Susswein LR, Zorn KK, Hiraki S, Arvai KJ, et al. Germline pathogenic variants identified in women with ovarian tumors. Gynecol Oncol. 2018;151(3):481–8. https://doi.org/10.1016/j.ygyno.2018.09.030.
26. Hirasawa A, Imoto I, Naruto T, Akahane T, Yamagami W, Nomura H, et al. Prevalence of pathogenic germline variants detected by multigene sequencing in unselected Japanese patients with ovarian cancer. Oncotarget. 2017;8(68):112258–67. https://doi.org/10.18632/oncotarget.22733.
27. Long B, Lilyquist J, Weaver A, Hu C, Gnanaolivu R, Lee KY, et al. Cancer susceptibility gene mutations in type I and II endometrial cancer. Gynecol Oncol. 2019;152(1):20–5. https://doi.org/10.1016/j.ygyno.2018.10.019.
28. Ring KL, Bruegl AS, Allen BA, Elkin EP, Singh N, Hartman AR, et al. Germline multigene hereditary cancer panel testing in an unselected endometrial cancer cohort. Mod Pathol. 2016;29(11):1381–9. https://doi.org/10.1038/modpathol.2016.135.
29. Susswein LR, Marshall ML, Nusbaum R, Vogel Postula KJ, Weissman SM, Yackowski L, et al. Pathogenic and likely pathogenic variant prevalence among the first 10,000 patients referred for next-generation cancer panel testing. Genet Med. 2016;18(8):823–32. https://doi.org/10.1038/gim.2015.166.
30. Kuchenbaecker KB, Hopper JL, Barnes DR, Phillips KA, Mooij TM, Roos-Blom MJ, et al. Risks of breast, ovarian, and contralateral breast cancer for BRCA1 and BRCA2 mutation carriers. J Am Med Assoc. 2017;317(23):2402–16. https://doi.org/10.1001/jama.2017.7112.
31. Zhang S, Royer R, Li S, McLaughlin JR, Rosen B, Risch HA, et al. Frequencies of BRCA1 and BRCA2 mutations among 1,342 unselected patients with invasive ovarian cancer. Gynecol Oncol. 2011;121(2):353–7. https://doi.org/10.1016/j.ygyno.2011.01.020.
32. Alsop K, Fereday S, Meldrum C, de Fazio A, Emmanuel C, George J, et al. BRCA mutation frequency and patterns of treatment response in BRCA mutation-positive women with ovarian cancer: a report from the Australian Ovarian Cancer Study Group. J Clin Oncol. 2012;30(21):2654–63. https://doi.org/10.1200/jco.2011.39.8545.
33. Bolton KL, Chenevix-Trench G, Goh C, Sadetzki S, Ramus SJ, Karlan BY, et al. Association between BRCA1 and BRCA2 mutations and survival in women with invasive epithelial ovarian cancer. J Am Med Assoc. 2012;307(4):382–90. https://doi.org/10.1001/jama.2012.20.
34. Yang D, Khan S, Sun Y, Hess K, Shmulevich I, Sood AK, et al. Association of BRCA1 and BRCA2 mutations with survival, chemotherapy sensitivity, and gene mutator phenotype in patients with ovarian cancer. J Am Med Assoc. 2011;306(14):1557–65. https://doi.org/10.1001/jama.2011.1456.
35. Segev Y, Iqbal J, Lubinski J, Gronwald J, Lynch HT, Moller P, et al. The incidence of endometrial cancer in women with BRCA1 and BRCA2 mutations: an international prospective cohort study. Gynecol Oncol. 2013;130(1):127–31. https://doi.org/10.1016/j.ygyno.2013.03.027.
36. Shu CA, Pike MC, Jotwani AR, Friebel TM, Soslow RA, Levine DA, et al. Uterine cancer after risk-reducing salpingo-oophorectomy without hysterectomy in women with BRCA mutations. JAMA Oncol. 2016;2(11):1434–40. https://doi.org/10.1001/jamaoncol.2016.1820.
37. Lynch HT, de la Chapelle A. Hereditary colorectal cancer. N Engl J Med. 2003;348(10):919–32. https://doi.org/10.1056/NEJMra012242.
38. Ligtenberg MJ, Kuiper RP, Chan TL, Goossens M, Hebeda KM, Voorendt M, et al. Heritable somatic methylation and inactivation of MSH2 in families with Lynch syndrome due to deletion of the 3′ exons of TACSTD1. Nat Genet. 2009;41(1):112–7. https://doi.org/10.1038/ng.283.

39. Dominguez-Valentin M, Sampson JR, Seppälä TT, Ten Broeke SW, Plazzer JP, Nakken S, et al. Cancer risks by gene, age, and gender in 6350 carriers of pathogenic mismatch repair variants: findings from the Prospective Lynch Syndrome Database. Genet Med. 2020;22(1):15–25. https://doi.org/10.1038/s41436-019-0596-9.
40. Rossi L, Le Frere-Belda MA, Laurent-Puig P, Buecher B, De Pauw A, Stoppa-Lyonnet D, et al. Clinicopathologic characteristics of endometrial cancer in Lynch syndrome: a French multicenter study. Int J Gynecol Cancer. 2017;27(5):953–60. https://doi.org/10.1097/igc.0000000000000985.
41. Helder-Woolderink JM, Blok EA, Vasen HF, Hollema H, Mourits MJ, De Bock GH. Ovarian cancer in Lynch syndrome; a systematic review. Eur J Cancer. 2016;55:65–73. https://doi.org/10.1016/j.ejca.2015.12.005.
42. Engel C, Loeffler M, Steinke V, Rahner N, Holinski-Feder E, Dietmaier W, et al. Risks of less common cancers in proven mutation carriers with lynch syndrome. J Clin Oncol. 2012;30(35):4409–15. https://doi.org/10.1200/jco.2012.43.2278.
43. Song H, Cicek MS, Dicks E, Harrington P, Ramus SJ, Cunningham JM, et al. The contribution of deleterious germline mutations in BRCA1, BRCA2 and the mismatch repair genes to ovarian cancer in the population. Hum Mol Genet. 2014;23(17):4703–9. https://doi.org/10.1093/hmg/ddu172.
44. Pal T, Akbari MR, Sun P, Lee JH, Fulp J, Thompson Z, et al. Frequency of mutations in mismatch repair genes in a population-based study of women with ovarian cancer. Br J Cancer. 2012;107(10):1783–90. https://doi.org/10.1038/bjc.2012.452.
45. Watson P, Bützow R, Lynch HT, Mecklin JP, Järvinen HJ, Vasen HF, et al. The clinical features of ovarian cancer in hereditary nonpolyposis colorectal cancer. Gynecol Oncol. 2001;82(2):223–8. https://doi.org/10.1006/gyno.2001.6279.
46. Nelen MR, Padberg GW, Peeters EA, Lin AY, van den Helm B, Frants RR, et al. Localization of the gene for Cowden disease to chromosome 10q22-23. Nat Genet. 1996;13(1):114–6. https://doi.org/10.1038/ng0596-114.
47. Tan MH, Mester JL, Ngeow J, Rybicki LA, Orloff MS, Eng C. Lifetime cancer risks in individuals with germline PTEN mutations. Clin Cancer Res. 2012;18(2):400–7. https://doi.org/10.1158/1078-0432.Ccr-11-2283.
48. Beggs AD, Latchford AR, Vasen HF, Moslein G, Alonso A, Aretz S, et al. Peutz-Jeghers syndrome: a systematic review and recommendations for management. Gut. 2010;59(7):975–86. https://doi.org/10.1136/gut.2009.198499.
49. Giardiello FM, Brensinger JD, Tersmette AC, Goodman SN, Petersen GM, Booker SV, et al. Very high risk of cancer in familial Peutz-Jeghers syndrome. Gastroenterology. 2000;119(6):1447–53. https://doi.org/10.1053/gast.2000.20228.
50. Jenne DE, Reimann H, Nezu J, Friedel W, Loff S, Jeschke R, et al. Peutz-Jeghers syndrome is caused by mutations in a novel serine threonine kinase. Nat Genet. 1998;18(1):38–43. https://doi.org/10.1038/ng0198-38.
51. Hemminki A, Markie D, Tomlinson I, Avizienyte E, Roth S, Loukola A, et al. A serine/threonine kinase gene defective in Peutz-Jeghers syndrome. Nature. 1998;391(6663):184–7. https://doi.org/10.1038/34432.
52. McGarrity TJ, Amos CI, Baker MJ. Peutz-Jeghers syndrome. In: Gene reviews. https://www.ncbi.nlm.nih.gov/books/NBK1266/
53. Young RH, Welch WR, Dickersin GR, Scully RE. Ovarian sex cord tumor with annular tubules: review of 74 cases including 27 with Peutz-Jeghers syndrome and four with adenoma malignum of the cervix. Cancer. 1982;50(7):1384–402. https://doi.org/10.1002/1097-0142(19821001)50:7<1384::aid-cncr2820500726>3.0.co;2-5.
54. Gilks CB, Young RH, Aguirre P, DeLellis RA, Scully RE. Adenoma malignum (minimal deviation adenocarcinoma) of the uterine cervix. A clinicopathological and immunohistochemical analysis of 26 cases. Am J Surg Pathol. 1989;13(9):717–29. https://doi.org/10.1097/00000478-198909000-00001.

55. Chen KT. Female genital tract tumors in Peutz-Jeghers syndrome. Hum Pathol. 1986;17(8):858–61. https://doi.org/10.1016/s0046-8177(86)80208-8.
56. Mikami Y, Kiyokawa T, Hata S, Fujiwara K, Moriya T, Sasano H, et al. Gastrointestinal immunophenotype in adenocarcinomas of the uterine cervix and related glandular lesions: a possible link between lobular endocervical glandular hyperplasia/pyloric gland metaplasia and 'adenoma malignum'. Mod Pathol. 2004;17(8):962–72. https://doi.org/10.1038/modpathol.3800148.
57. Hirasawa A, Akahane T, Tsuruta T, Kobayashi Y, Masuda K, Banno K, et al. Lobular endocervical glandular hyperplasia and peritoneal pigmentation associated with Peutz-Jeghers syndrome due to a germline mutation of STK11. Ann Oncol. 2012;23(11):2990–2. https://doi.org/10.1093/annonc/mds492.
58. Ito M, Minamiguchi S, Mikami Y, Ueda Y, Sekiyama K, Yamamoto T, et al. Peutz-Jeghers syndrome-associated atypical mucinous proliferation of the uterine cervix: a case of minimal deviation adenocarcinoma ('adenoma malignum') in situ. Pathol Res Pract. 2012;208(10):623–7. https://doi.org/10.1016/j.prp.2012.06.008.
59. Kobayashi Y, Masuda K, Kimura T, Nomura H, Hirasawa A, Banno K, et al. A tumor of the uterine cervix with a complex histology in a Peutz-Jeghers syndrome patient with genomic deletion of the STK11 exon 1 region. Future Oncol. 2014;10(2):171–7. https://doi.org/10.2217/fon.13.180.
60. Takei Y, Fujiwara H, Nagashima T, Takahashi Y, Takahashi S, Suzuki M. Successful pregnancy in a Peutz-Jeghers syndrome patient with lobular endocervical glandular hyperplasia. J Obstet Gynaecol Res. 2015;41(3):468–73. https://doi.org/10.1111/jog.12541.
61. Slade I, Bacchelli C, Davies H, Murray A, Abbaszadeh F, Hanks S, et al. DICER1 syndrome: clarifying the diagnosis, clinical features and management implications of a pleiotropic tumour predisposition syndrome. J Med Genet. 2011;48(4):273–8. https://doi.org/10.1136/jmg.2010.083790.
62. Schutlz KA, Stewart DR, Kamihara J, Bauer AJ, Merideth MA, Stratton P, et al. *DICER1* tumor predisposition. In Gene reviews. https://www.ncbi.nlm.nih.gov/books/NBK196157/
63. Hill DA, Ivanovich J, Priest JR, Gurnett CA, Dehner LP, Desruisseau D, et al. DICER1 mutations in familial pleuropulmonary blastoma. Science. 2009;325(5943):965. https://doi.org/10.1126/science.1174334.
64. Foulkes WD, Priest JR, Duchaine TF. DICER1: mutations, microRNAs and mechanisms. Nat Rev Cancer. 2014;14(10):662–72. https://doi.org/10.1038/nrc3802.
65. Brenneman M, Field A, Yang J, Williams G, Doros L, Rossi C, et al. Temporal order of RNase IIIb and loss-of-function mutations during development determines phenotype in pleuropulmonary blastoma/DICER1 syndrome: a unique variant of the two-hit tumor suppression model. F1000Res. 2015;4:214. https://doi.org/10.12688/f1000research.6746.2.
66. Lambertz I, Nittner D, Mestdagh P, Denecker G, Vandesompele J, Dyer MA, et al. Monoallelic but not biallelic loss of Dicer1 promotes tumorigenesis in vivo. Cell Death Differ. 2010;17(4):633–41. https://doi.org/10.1038/cdd.2009.202.
67. Schultz KAP, Williams GM, Kamihara J, Stewart DR, Harris AK, Bauer AJ, et al. DICER1 and associated conditions: identification of at-risk individuals and recommended surveillance strategies. Clin Cancer Res. 2018;24(10):2251–61. https://doi.org/10.1158/1078-0432.Ccr-17-3089.
68. Stewart DR, Best AF, Williams GM, Harney LA, Carr AG, Harris AK, et al. Neoplasm risk among individuals with a pathogenic germline variant in DICER1. J Clin Oncol. 2019;37(8):668–76. https://doi.org/10.1200/jco.2018.78.4678.
69. de Kock L, Terzic T, McCluggage WG, Stewart CJR, Shaw P, Foulkes WD, et al. DICER1 mutations are consistently present in moderately and poorly differentiated Sertoli-Leydig cell tumors. Am J Surg Pathol. 2017;41(9):1178–87. https://doi.org/10.1097/pas.0000000000000895.
70. Minard-Colin V, Walterhouse D, Bisogno G, Martelli H, Anderson J, Rodeberg DA, et al. Localized vaginal/uterine rhabdomyosarcoma-results of a pooled analysis from four interna-

tional cooperative groups. Pediatr Blood Cancer. 2018;65(9):e27096. https://doi.org/10.1002/pbc.27096.

71. Dehner LP, Jarzembowski JA, Hill DA. Embryonal rhabdomyosarcoma of the uterine cervix: a report of 14 cases and a discussion of its unusual clinicopathological associations. Mod Pathol. 2012;25(4):602–14. https://doi.org/10.1038/modpathol.2011.185.
72. de Kock L, Yoon JY, Apellaniz-Ruiz M, Pelletier D, McCluggage WG, Stewart CJR, et al. Significantly greater prevalence of DICER1 alterations in uterine embryonal rhabdomyosarcoma compared to adenosarcoma. Mod Pathol. 2020;33(6):1207–19. https://doi.org/10.1038/s41379-019-0436-0.
73. Jelinic P, Mueller JJ, Olvera N, Dao F, Scott SN, Shah R, et al. Recurrent SMARCA4 mutations in small cell carcinoma of the ovary. Nat Genet. 2014;46(5):424–6. https://doi.org/10.1038/ng.2922.
74. Ramos P, Karnezis AN, Craig DW, Sekulic A, Russell ML, Hendricks WP, et al. Small cell carcinoma of the ovary, hypercalcemic type, displays frequent inactivating germline and somatic mutations in SMARCA4. Nat Genet. 2014;46(5):427–9. https://doi.org/10.1038/ng.2928.
75. Witkowski L, Carrot-Zhang J, Albrecht S, Fahiminiya S, Hamel N, Tomiak E, et al. Germline and somatic SMARCA4 mutations characterize small cell carcinoma of the ovary, hypercalcemic type. Nat Genet. 2014;46(5):438–43. https://doi.org/10.1038/ng.2931.
76. Witkowski L, Goudie C, Ramos P, Boshari T, Brunet JS, Karnezis AN, et al. The influence of clinical and genetic factors on patient outcome in small cell carcinoma of the ovary, hypercalcemic type. Gynecol Oncol. 2016;141(3):454–60. https://doi.org/10.1016/j.ygyno.2016.03.013.
77. Sredni ST, Tomita T. Rhabdoid tumor predisposition syndrome. Pediatr Dev Pathol. 2015;18(1):49–58. https://doi.org/10.2350/14-07-1531-misc.1.
78. Tischkowitz M, Huang S, Banerjee S, Hague J, Hendricks WPD, Huntsman DG, et al. Small-cell carcinoma of the ovary, hypercalcemic type-genetics, new treatment targets, and current management guidelines. Clin Cancer Res. 2020;26(15):3908–17. https://doi.org/10.1158/1078-0432.Ccr-19-3797.
79. Kurian AW, Ward KC, Howlader N, Deapen D, Hamilton AS, Mariotto A, et al. Genetic testing and results in a population-based cohort of breast cancer patients and ovarian cancer patients. J Clin Oncol. 2019;37(15):1305–15. https://doi.org/10.1200/jco.18.01854.
80. Suszynska M, Klonowska K, Jasinska AJ, Kozlowski P. Large-scale meta-analysis of mutations identified in panels of breast/ovarian cancer-related genes—providing evidence of cancer predisposition genes. Gynecol Oncol. 2019;153(2):452–62. https://doi.org/10.1016/j.ygyno.2019.01.027.
81. Ramus SJ, Song H, Dicks E, Tyrer JP, Rosenthal AN, Intermaggio MP, et al. Germline mutations in the BRIP1, BARD1, PALB2, and NBN genes in women with ovarian cancer. J Natl Cancer Inst. 2015;107(11):djv214. https://doi.org/10.1093/jnci/djv214.
82. Lilyquist J, LaDuca H, Polley E, Davis BT, Shimelis H, Hu C, et al. Frequency of mutations in a large series of clinically ascertained ovarian cancer cases tested on multi-gene panels compared to reference controls. Gynecol Oncol. 2017;147(2):375–80. https://doi.org/10.1016/j.ygyno.2017.08.030.
83. Kurian AW, Hughes E, Handorf EA, Gutin A, Allen B, Hartman AR, et al. Breast and ovarian cancer penetrance estimates derived from germline multiple-gene sequencing results in women. JCO Precis Oncol. 2017;1:1–12. https://doi.org/10.1200/po.16.00066.
84. Antoniou AC, Casadei S, Heikkinen T, Barrowdale D, Pylkäs K, Roberts J, et al. Breast-cancer risk in families with mutations in PALB2. N Engl J Med. 2014;371(6):497–506. https://doi.org/10.1056/NEJMoa1400382.
85. Song H, Dicks E, Ramus SJ, Tyrer JP, Intermaggio MP, Hayward J, et al. Contribution of germline mutations in the RAD51B, RAD51C, and RAD51D genes to ovarian cancer in the population. J Clin Oncol. 2015;33(26):2901–7. https://doi.org/10.1200/jco.2015.61.2408.
86. Hampel H, Bennett RL, Buchanan A, Pearlman R, Wiesner GL. A practice guideline from the American College of Medical Genetics and Genomics and the National Society of

Genetic Counselors: referral indications for cancer predisposition assessment. Genet Med. 2015;17(1):70–87. https://doi.org/10.1038/gim.2014.147.

87. Lancaster JM, Powell CB, Chen LM, Richardson DL. Society of Gynecologic Oncology statement on risk assessment for inherited gynecologic cancer predispositions. Gynecol Oncol. 2015;136(1):3–7. https://doi.org/10.1016/j.ygyno.2014.09.009.
88. Owens DK, Davidson KW, Krist AH, Barry MJ, Cabana M, Caughey AB, et al. Risk assessment, genetic counseling, and genetic testing for BRCA-related cancer: US Preventive Services Task Force recommendation statement. J Am Med Assoc. 2019;322(7):652–65. https://doi.org/10.1001/jama.2019.10987.
89. Robson ME, Bradbury AR, Arun B, Domchek SM, Ford JM, Hampel HL, et al. American Society of Clinical Oncology policy statement update: genetic and genomic testing for cancer susceptibility. J Clin Oncol. 2015;33(31):3660–7. https://doi.org/10.1200/jco.2015.63.0996.
90. Practice bulletin no182: hereditary breast and ovarian cancer syndrome. Obstet Gynecol. 2017;130(3):e110–26. https://doi.org/10.1097/AOG.0000000000002296.
91. Parmigiani G, Berry D, Aguilar O. Determining carrier probabilities for breast cancer-susceptibility genes BRCA1 and BRCA2. Am J Hum Genet. 1998;62(1):145–58. https://doi.org/10.1086/301670.
92. Antoniou AC, Pharoah PP, Smith P, Easton DF. The BOADICEA model of genetic susceptibility to breast and ovarian cancer. Br J Cancer. 2004;91(8):1580–90. https://doi.org/10.1038/sj.bjc.6602175.
93. Kastrinos F, Uno H, Ukaegbu C, Alvero C, McFarland A, Yurgelun MB, et al. Development and validation of the PREMM(5) model for comprehensive risk assessment of Lynch syndrome. J Clin Oncol. 2017;35(19):2165–72. https://doi.org/10.1200/jco.2016.69.6120.
94. Barnetson RA, Tenesa A, Farrington SM, Nicholl ID, Cetnarskyj R, Porteous ME, et al. Identification and survival of carriers of mutations in DNA mismatch-repair genes in colon cancer. N Engl J Med. 2006;354(26):2751–63. https://doi.org/10.1056/NEJMoa053493.
95. Chen S, Wang W, Lee S, Nafa K, Lee J, Romans K, et al. Prediction of germline mutations and cancer risk in the Lynch syndrome. J Am Med Assoc. 2006;296(12):1479–87. https://doi.org/10.1001/jama.296.12.1479.
96. Vogel KJ, Atchley DP, Erlichman J, Broglio KR, Ready KJ, Valero V, et al. BRCA1 and BRCA2 genetic testing in Hispanic patients: mutation prevalence and evaluation of the BRCAPRO risk assessment model. J Clin Oncol. 2007;25(29):4635–41. https://doi.org/10.1200/jco.2006.10.4703.
97. Kurian AW, Gong GD, John EM, Miron A, Felberg A, Phipps AI, et al. Performance of prediction models for BRCA mutation carriage in three racial/ethnic groups: findings from the Northern California Breast Cancer Family Registry. Cancer Epidemiol Biomark Prev. 2009;18(4):1084–91. https://doi.org/10.1158/1055-9965.Epi-08-1090.
98. Kurian AW, Gong GD, Chun NM, Mills MA, Staton AD, Kingham KE, et al. Performance of BRCA1/2 mutation prediction models in Asian Americans. J Clin Oncol. 2008;26(29):4752–8. https://doi.org/10.1200/jco.2008.16.8310.
99. Biswas S, Tankhiwale N, Blackford A, Barrera AM, Ready K, Lu K, et al. Assessing the added value of breast tumor markers in genetic risk prediction model BRCAPRO. Breast Cancer Res Treat. 2012;133(1):347–55. https://doi.org/10.1007/s10549-012-1958-z.
100. Ready KJ, Vogel KJ, Atchley DP, Broglio KR, Solomon KK, Amos C, et al. Accuracy of the BRCAPRO model among women with bilateral breast cancer. Cancer. 2009;115(4):725–30. https://doi.org/10.1002/cncr.24102.
101. Daniels MS, Babb SA, King RH, Urbauer DL, Batte BA, Brandt AC, et al. Underestimation of risk of a BRCA1 or BRCA2 mutation in women with high-grade serous ovarian cancer by BRCAPRO: a multi-institution study. J Clin Oncol. 2014;32(12):1249–55. https://doi.org/10.1200/jco.2013.50.6055.
102. Latham A, Srinivasan P, Kemel Y, Shia J, Bandlamudi C, Mandelker D, et al. Microsatellite instability is associated with the presence of Lynch syndrome pan-cancer. J Clin Oncol. 2019;37(4):286–95. https://doi.org/10.1200/jco.18.00283.

103. Mandelker D, Donoghue M, Talukdar S, Bandlamudi C, Srinivasan P, Vivek M, et al. Germline-focussed analysis of tumour-only sequencing: recommendations from the ESMO Precision Medicine Working Group. Ann Oncol. 2019;30(8):1221–31. https://doi.org/10.1093/annonc/mdz136.
104. Tandy-Connor S, Guiltinan J, Krempely K, LaDuca H, Reineke P, Gutierrez S, et al. False-positive results released by direct-to-consumer genetic tests highlight the importance of clinical confirmation testing for appropriate patient care. Genet Med. 2018;20(12):1515–21. https://doi.org/10.1038/gim.2018.38.
105. Beitsch PD, Whitworth PW, Hughes K, Patel R, Rosen B, Compagnoni G, et al. underdiagnosis of hereditary breast cancer: are genetic testing guidelines a tool or an obstacle? J Clin Oncol. 2019;37(6):453–60. https://doi.org/10.1200/jco.18.01631.
106. Yadav S, Hu C, Hart SN, Boddicker N, Polley EC, Na J, et al. Evaluation of germline genetic testing criteria in a hospital-based series of women with breast cancer. J Clin Oncol. 2020;38(13):1409–18. https://doi.org/10.1200/jco.19.02190.
107. Ward M, Elder B, Habtemariam M. Current testing guidelines: a retrospective analysis of a community-based hereditary cancer program. J Adv Pract Oncol. 2021;12(7):693–701. https://doi.org/10.6004/jadpro.2021.12.7.3.
108. Kurian AW, Ward KC, Hamilton AS, Deapen DM, Abrahamse P, Bondarenko I, et al. Uptake, results, and outcomes of germline multiple-gene sequencing after diagnosis of breast cancer. JAMA Oncol. 2018;4(8):1066–72. https://doi.org/10.1001/jamaoncol.2018.0644.
109. Resta R, Biesecker BB, Bennett RL, Blum S, Hahn SE, Strecker MN, et al. A new definition of Genetic Counseling: National Society of Genetic Counselors' Task Force report. J Genet Couns. 2006;15(2):77–83. https://doi.org/10.1007/s10897-005-9014-3.
110. Kast K, Rhiem K, Wappenschmidt B, Hahnen E, Hauke J, Bluemcke B, et al. Prevalence of BRCA1/2 germline mutations in 21 401 families with breast and ovarian cancer. J Med Genet. 2016;53(7):465–71. https://doi.org/10.1136/jmedgenet-2015-103672.
111. Passaperuma K, Warner E, Causer PA, Hill KA, Messner S, Wong JW, et al. Long-term results of screening with magnetic resonance imaging in women with BRCA mutations. Br J Cancer. 2012;107(1):24–30. https://doi.org/10.1038/bjc.2012.204.
112. Lehman CD, Lee JM, DeMartini WB, Hippe DS, Rendi MH, Kalish G, et al. Screening MRI in women with a personal history of breast cancer. J Natl Cancer Inst. 2016;108(3):djv349. https://doi.org/10.1093/jnci/djv349.
113. Konstantinopoulos PA, Norquist B, Lacchetti C, Armstrong D, Grisham RN, Goodfellow PJ, et al. Germline and somatic tumor testing in epithelial ovarian cancer: ASCO guideline. J Clin Oncol. 2020;38(11):1222–45. https://doi.org/10.1200/jco.19.02960.
114. Paluch-Shimon S, Cardoso F, Sessa C, Balmana J, Cardoso MJ, Gilbert F, et al. Prevention and screening in BRCA mutation carriers and other breast/ovarian hereditary cancer syndromes: ESMO Clinical Practice Guidelines for cancer prevention and screening. Ann Oncol. 2016;27(suppl 5):v103–v10. https://doi.org/10.1093/annonc/mdw327.
115. Finch AP, Lubinski J, Møller P, Singer CF, Karlan B, Senter L, et al. Impact of oophorectomy on cancer incidence and mortality in women with a BRCA1 or BRCA2 mutation. J Clin Oncol. 2014;32(15):1547–53. https://doi.org/10.1200/jco.2013.53.2820.
116. Rebbeck TR, Kauff ND, Domchek SM. Meta-analysis of risk reduction estimates associated with risk-reducing salpingo-oophorectomy in BRCA1 or BRCA2 mutation carriers. J Natl Cancer Inst. 2009;101(2):80–7. https://doi.org/10.1093/jnci/djn442.
117. Kauff ND, Domchek SM, Friebel TM, Robson ME, Lee J, Garber JE, et al. Risk-reducing salpingo-oophorectomy for the prevention of BRCA1- and BRCA2-associated breast and gynecologic cancer: a multicenter, prospective study. J Clin Oncol. 2008;26(8):1331–7. https://doi.org/10.1200/jco.2007.13.9626.
118. Callahan MJ, Crum CP, Medeiros F, Kindelberger DW, Elvin JA, Garber JE, et al. Primary fallopian tube malignancies in BRCA-positive women undergoing surgery for ovarian cancer risk reduction. J Clin Oncol. 2007;25(25):3985–90. https://doi.org/10.1200/jco.2007.12.2622.

119. Powell CB, Kenley E, Chen LM, Crawford B, McLennan J, Zaloudek C, et al. Risk-reducing salpingo-oophorectomy in BRCA mutation carriers: role of serial sectioning in the detection of occult malignancy. J Clin Oncol. 2005;23(1):127–32. https://doi.org/10.1200/jco.2005.04.109.
120. Finch A, Shaw P, Rosen B, Murphy J, Narod SA, Colgan TJ. Clinical and pathologic findings of prophylactic salpingo-oophorectomies in 159 BRCA1 and BRCA2 carriers. Gynecol Oncol. 2006;100(1):58–64. https://doi.org/10.1016/j.ygyno.2005.06.065.
121. Harmsen MG, Piek JMJ, Bulten J, Casey MJ, Rebbeck TR, Mourits MJ, et al. Peritoneal carcinomatosis after risk-reducing surgery in BRCA1/2 mutation carriers. Cancer. 2018;124(5):952–9. https://doi.org/10.1002/cncr.31211.
122. Marchetti C, De Felice F, Boccia S, Sassu C, Di Donato V, Perniola G, et al. Hormone replacement therapy after prophylactic risk-reducing salpingo-oophorectomy and breast cancer risk in BRCA1 and BRCA2 mutation carriers: a meta-analysis. Crit Rev Oncol Hematol. 2018;132:111–5. https://doi.org/10.1016/j.critrevonc.2018.09.018.
123. Gordhandas S, Norquist BM, Pennington KP, Yung RL, Laya MB, Swisher EM. Hormone replacement therapy after risk reducing salpingo-oophorectomy in patients with BRCA1 or BRCA2 mutations; a systematic review of risks and benefits. Gynecol Oncol. 2019;153(1):192–200. https://doi.org/10.1016/j.ygyno.2018.12.014.
124. Kotsopoulos J, Huzarski T, Gronwald J, Moller P, Lynch HT, Neuhausen SL, et al. Hormone replacement therapy after menopause and risk of breast cancer in BRCA1 mutation carriers: a case-control study. Breast Cancer Res Treat. 2016;155(2):365–73. https://doi.org/10.1007/s10549-016-3685-3.
125. Chlebowski RT, Anderson GL, Aragaki AK, Manson JE, Stefanick ML, Pan K, et al. Association of menopausal hormone therapy with breast cancer incidence and mortality during long-term follow-up of the women's health initiative randomized clinical trials. J Am Med Assoc. 2020;324(4):369–80. https://doi.org/10.1001/jama.2020.9482.
126. Harmsen MG, IntHout J, Arts-de Jong M, Hoogerbrugge N, Massuger L, Hermens R, et al. Salpingectomy with delayed oophorectomy in BRCA1/2 mutation carriers: estimating ovarian cancer risk. Obstet Gynecol. 2016;127(6):1054–63. https://doi.org/10.1097/aog.0000000000001448.
127. Daly MB, Dresher CW, Yates MS, Jeter JM, Karlan BY, Alberts DS, et al. Salpingectomy as a means to reduce ovarian cancer risk. Cancer Prev Res (Phila). 2015;8(5):342–8. https://doi.org/10.1158/1940-6207.Capr-14-0293.
128. Beral V, Doll R, Hermon C, Peto R, Reeves G. Ovarian cancer and oral contraceptives: collaborative reanalysis of data from 45 epidemiological studies including 23,257 women with ovarian cancer and 87,303 controls. Lancet. 2008;371(9609):303–14. https://doi.org/10.1016/s0140-6736(08)60167-1.
129. Iodice S, Barile M, Rotmensz N, Feroce I, Bonanni B, Radice P, et al. Oral contraceptive use and breast or ovarian cancer risk in BRCA1/2 carriers: a meta-analysis. Eur J Cancer. 2010;46(12):2275–84. https://doi.org/10.1016/j.ejca.2010.04.018.
130. Cibula D, Zikan M, Dusek L, Majek O. Oral contraceptives and risk of ovarian and breast cancers in BRCA mutation carriers: a meta-analysis. Expert Rev Anticancer Ther. 2011;11(8):1197–207. https://doi.org/10.1586/era.11.38.
131. Moorman PG, Havrilesky LJ, Gierisch JM, Coeytaux RR, Lowery WJ, Peragallo Urrutia R, et al. Oral contraceptives and risk of ovarian cancer and breast cancer among high-risk women: a systematic review and meta-analysis. J Clin Oncol. 2013;31(33):4188–98. https://doi.org/10.1200/jco.2013.48.9021.
132. Narod SA, Dubé MP, Klijn J, Lubinski J, Lynch HT, Ghadirian P, et al. Oral contraceptives and the risk of breast cancer in BRCA1 and BRCA2 mutation carriers. J Natl Cancer Inst. 2002;94(23):1773–9. https://doi.org/10.1093/jnci/94.23.1773.
133. Haile RW, Thomas DC, McGuire V, Felberg A, John EM, Milne RL, et al. BRCA1 and BRCA2 mutation carriers, oral contraceptive use, and breast cancer before age 50. Cancer Epidemiol Biomark Prev. 2006;15(10):1863–70. https://doi.org/10.1158/1055-9965.Epi-06-0258.

134. Milne RL, Knight JA, John EM, Dite GS, Balbuena R, Ziogas A, et al. Oral contraceptive use and risk of early-onset breast cancer in carriers and noncarriers of BRCA1 and BRCA2 mutations. Cancer Epidemiol Biomark Prev. 2005;14(2):350–6. https://doi.org/10.1158/1055-9965.Epi-04-0376.
135. Lee E, Ma H, McKean-Cowdin R, Van Den Berg D, Bernstein L, Henderson BE, et al. Effect of reproductive factors and oral contraceptives on breast cancer risk in BRCA1/2 mutation carriers and noncarriers: results from a population-based study. Cancer Epidemiol Biomark Prev. 2008;17(11):3170–8. https://doi.org/10.1158/1055-9965.Epi-08-0396.
136. Ding YC, Steele L, Kuan CJ, Greilac S, Neuhausen SL. Mutations in BRCA2 and PALB2 in male breast cancer cases from the United States. Breast Cancer Res Treat. 2011;126(3):771–8. https://doi.org/10.1007/s10549-010-1195-2.
137. Friedman LS, Gayther SA, Kurosaki T, Gordon D, Noble B, Casey G, et al. Mutation analysis of BRCA1 and BRCA2 in a male breast cancer population. Am J Hum Genet. 1997;60(2):313–9.
138. Evans DG, Susnerwala I, Dawson J, Woodward E, Maher ER, Lalloo F. Risk of breast cancer in male BRCA2 carriers. J Med Genet. 2010;47(10):710–1. https://doi.org/10.1136/jmg.2009.075176.
139. Tai YC, Domchek S, Parmigiani G, Chen S. Breast cancer risk among male BRCA1 and BRCA2 mutation carriers. J Natl Cancer Inst. 2007;99(23):1811–4. https://doi.org/10.1093/jnci/djm203.
140. Leongamornlert D, Mahmud N, Tymrakiewicz M, Saunders E, Dadaev T, Castro E, et al. Germline BRCA1 mutations increase prostate cancer risk. Br J Cancer. 2012;106(10):1697–701. https://doi.org/10.1038/bjc.2012.146.
141. Abida W, Armenia J, Gopalan A, Brennan R, Walsh M, Barron D, et al. Prospective genomic profiling of prostate cancer across disease states reveals germline and somatic alterations that may affect clinical decision making. JCO Precis Oncol. 2017;2017:PO.17.00029. https://doi.org/10.1200/po.17.00029.
142. Giri VN, Hegarty SE, Hyatt C, O'Leary E, Garcia J, Knudsen KE, et al. Germline genetic testing for inherited prostate cancer in practice: implications for genetic testing, precision therapy, and cascade testing. Prostate. 2019;79(4):333–9. https://doi.org/10.1002/pros.23739.
143. Lang SH, Swift SL, White H, Misso K, Kleijnen J, Quek RGW. A systematic review of the prevalence of DNA damage response gene mutations in prostate cancer. Int J Oncol. 2019;55(3):597–616. https://doi.org/10.3892/ijo.2019.4842.
144. Goggins M, Overbeek KA, Brand R, Syngal S, Del Chiaro M, Bartsch DK, et al. Management of patients with increased risk for familial pancreatic cancer: updated recommendations from the International Cancer of the Pancreas Screening (CAPS) Consortium. Gut. 2020;69(1):7–17. https://doi.org/10.1136/gutjnl-2019-319352.
145. Canto MI, Almario JA, Schulick RD, Yeo CJ, Klein A, Blackford A, et al. Risk of neoplastic progression in individuals at high risk for pancreatic cancer undergoing long-term surveillance. Gastroenterology. 2018;155(3):740–51.e2. https://doi.org/10.1053/j.gastro.2018.05.035.
146. Vasen H, Ibrahim I, Ponce CG, Slater EP, Matthäi E, Carrato A, et al. Benefit of surveillance for pancreatic cancer in high-risk individuals: outcome of long-term prospective follow-up studies from three European expert centers. J Clin Oncol. 2016;34(17):2010–9. https://doi.org/10.1200/jco.2015.64.0730.
147. Balmaña J, Balaguer F, Cervantes A, Arnold D. Familial risk-colorectal cancer: ESMO Clinical Practice Guidelines. Ann Oncol. 2013;24(Suppl 6):vi73–80. https://doi.org/10.1093/annonc/mdt209.
148. Giardiello FM, Allen JI, Axilbund JE, Boland CR, Burke CA, Burt RW, et al. Guidelines on genetic evaluation and management of Lynch syndrome: a consensus statement by the US Multi-Society Task Force on colorectal cancer. Gastroenterology. 2014;147(2):502–26. https://doi.org/10.1053/j.gastro.2014.04.001.

149. Stoffel EM, Mangu PB, Gruber SB, Hamilton SR, Kalady MF, Lau MW, et al. Hereditary colorectal cancer syndromes: American Society of Clinical Oncology Clinical Practice Guideline endorsement of the familial risk-colorectal cancer: European Society for Medical Oncology Clinical Practice Guidelines. J Clin Oncol. 2015;33(2):209–17. https://doi.org/10.1200/jco.2014.58.1322.
150. Rubenstein JH, Enns R, Heidelbaugh J, Barkun A. American Gastroenterological Association Institute Guideline on the diagnosis and management of Lynch syndrome. Gastroenterology. 2015;149(3):777–82; quiz e16-7. https://doi.org/10.1053/j.gastro.2015.07.036.
151. Lindor NM, Petersen GM, Hadley DW, Kinney AY, Miesfeldt S, Lu KH, et al. Recommendations for the care of individuals with an inherited predisposition to Lynch syndrome: a systematic review. J Am Med Assoc. 2006;296(12):1507–17. https://doi.org/10.1001/jama.296.12.1507.
152. Syngal S, Brand RE, Church JM, Giardiello FM, Hampel HL, Burt RW. ACG clinical guideline: genetic testing and management of hereditary gastrointestinal cancer syndromes. Am J Gastroenterol. 2015;110(2):223–62; quiz 63. https://doi.org/10.1038/ajg.2014.435.
153. Tzortzatos G, Andersson E, Soller M, Askmalm MS, Zagoras T, Georgii-Hemming P, et al. The gynecological surveillance of women with Lynch syndrome in Sweden. Gynecol Oncol. 2015;138(3):717–22. https://doi.org/10.1016/j.ygyno.2015.07.016.
154. Gerritzen LH, Hoogerbrugge N, Oei AL, Nagengast FM, van Ham MA, Massuger LF, et al. Improvement of endometrial biopsy over transvaginal ultrasound alone for endometrial surveillance in women with Lynch syndrome. Familial Cancer. 2009;8(4):391–7. https://doi.org/10.1007/s10689-009-9252-x.
155. ACOG practive bulletin no.147: Lynch syndrome. Obstet Gynecol. 2014;124(5):1042–54. https://doi.org/10.1097/01.ACOG.0000456435.50739.72.
156. Schmeler KM, Lynch HT, Chen LM, Munsell MF, Soliman PT, Clark MB, et al. Prophylactic surgery to reduce the risk of gynecologic cancers in the Lynch syndrome. N Engl J Med. 2006;354(3):261–9. https://doi.org/10.1056/NEJMoa052627.
157. Bancroft EK, Page EC, Brook MN, Thomas S, Taylor N, Pope J, et al. A prospective prostate cancer screening program for men with pathogenic variants in mismatch repair genes (IMPACT): initial results from an international prospective study. Lancet Oncol. 2021;22(11):1618–31. https://doi.org/10.1016/s1470-2045(21)00522-2.
158. Bubien V, Bonnet F, Brouste V, Hoppe S, Barouk-Simonet E, David A, et al. High cumulative risks of cancer in patients with PTEN hamartoma tumour syndrome. J Med Genet. 2013;50(4):255–63. https://doi.org/10.1136/jmedgenet-2012-101339.
159. Riegert-Johnson DL, Gleeson FC, Roberts M, Tholen K, Youngborg L, Bullock M, et al. Cancer and Lhermitte-Duclos disease are common in Cowden syndrome patients. Hered Cancer Clin Pract. 2010;8(1):6. https://doi.org/10.1186/1897-4287-8-6.
160. Schultz KAP, Rednam SP, Kamihara J, Doros L, Achatz MI, Wasserman JD, et al. PTEN, DICER1, FH, and their associated tumor susceptibility syndromes: clinical features, genetics, and surveillance recommendations in childhood. Clin Cancer Res. 2017;23(12):e76–82. https://doi.org/10.1158/1078-0432.Ccr-17-0629.
161. Takatsu A, Shiozawa T, Miyamoto T, Kurosawa K, Kashima H, Yamada T, Kaku T, Mikami Y, Kiyokawa T, Tsuda H, Ishii K, Togashi K, Koyama T, Fujinaga Y, Kadoya M, Hashi A, Susumu N, Konishi I. Preoperative differential diagnosis of minimal deviation adenocarcinoma and lobular endocervical glandular hyperplasia of the uterine cervix. Int J Gynecol Cancer. 2011;21(7):1287–96. https://doi.org/10.1097/IGC.0b013e31821f746c.

# Chapter 8
# How Genome-Wide Analysis Contributes to Personalized Treatment in Cancer, Including Gynecologic Cancer?

**Hisamitsu Takaya**

**Abstract** Cancer omics analysis, which started with large-scale cancer genome data analysis, is becoming in this era multiomics analysis, which integrates multiple omics analyses with the development of analytical technologies. Omics analysis is being conducted in many parts of the world, the accumulated analysis data are rapidly growing, and new clinical trials are often conducted based on omics analysis. It is anticipated that future cancer treatments will require skills to search and analyze omics data more efficiently. In addition, different approaches are being used to validate data obtained from omics analyses in clinical practice compared to those used in the past. Master protocols are protocols designed with multiple subtrials within the framework of an overall trial structure, and they represent a paradigm shift in clinical trials, such as biomarker-driven clinical trials across cancer types or adaptive designs that allow for the interruption and addition of new subtrials. They are expected to play a role in the development of personalized treatment, which will become even more individualized in the future.

**Keywords** Cancer genome · Omics · Precision medicine · Master protocol · Umbrella trial · Basket trial · Platform trial

## 8.1 Introduction

Large-scale cancer genome analysis, which began with The Cancer Genome Atlas (TCGA), has spread widely as next-generation sequencers (NGS) have become more widely available, research costs have decreased, and research has been conducted into a variety of cancer types. As a result, not only genome analysis but also

H. Takaya (✉)
Department of Obstetrics and Gynecology, Kindai University, Osaka, Japan
e-mail: htakaya@med.kindai.ac.jp



M. Mandai (ed.), *Personalization in Gynecologic Oncology*, Comprehensive Gynecology and Obstetrics, https://doi.org/10.1007/978-981-19-4711-7_8

various omics analyses, such as gene expression, protein, and metabolite analyses, have been conducted, and knowledge of cancer omics has become indispensable to current cancer research. Drug discovery and clinical trials using these omics data are also being conducted, and knowledge in this field is becoming indispensable for clinicians.

In this chapter, we introduce the basic knowledge about omics data, databases required for actual handling of omics data, and analysis methods for actual cancer genome data using mainly TCGA data to understand personalized therapy using cancer genome data. Clinical trials using the new technology are also discussed, as well as the current status of clinical trials in gynecological oncology.

## 8.2 Omics Analysis

The word "genome" was coined by H. Winkler in 1920 [1] as "the set of chromosomes carried by gametes" and later redefined by Hitoshi Kihara [2] as "the minimum set of chromosomes essential to make an organism what it is." It was coined using the Greek suffix "-ome," meaning "all." Subsequently, as the analysis of genomes progressed, analysis to grasp the entire picture of mRNA and proteins, which are gene transcripts, as well as metabolic products, was promoted, giving rise to the terms "transcriptome," "proteome," and "metabolome," with "-ome" as a derivative of genome. In addition to molecular information, there are many other terms with "-ome," such as interactome, which is comprehensive information on interactions between molecules in living organisms, and microbiome, which is comprehensive information about bacterial flora. The term "omics" refers to the field of research that addresses these "-omes"; the comprehensive analysis of each is collectively called "omics analysis," and the information obtained by the analysis is called "omics information."

Cancer is a disease that encompasses an extremely complex system. As a system, cancer cells are intricately involved in interactions with surrounding tissues, such as the microenvironment and immune system, interactions between tumor cells, and factors such as transcriptional regulation, gene coexpression, signal transduction, metabolic pathways, and protein interactions within tumor cells; these layers of factors must be elucidated to understand the phenomenon of cancer as a disease [3, 4]. Therefore, a method called multiomics analysis, which integrates omics information involving cancer, is now being used to characterize the cancer system at a phenomenological level [5, 6]. The main omics analyses used in multiomics analysis include genomics, which addresses genomic sequences and their mutations, such as insertions, deletions, single nucleotide variations, and copy number variations; epigenomics, which analyzes DNA methylation, histone modifications, chromatin accessibility, and chromosomal 3D structure; transcriptomics, which analyzes quantitative gene expression and measures transcripts, such as microRNAs and long noncoding RNAs; proteomics, which analyzes protein expression and quantification, posttranslational modifications and protein–protein interactions; and metabolomics, which analyzes the quantification of metabolites of small molecules, such as amino acids, fatty acids, and carbohydrates. NGS is mainly used for genomics, epigenomics, and transcriptomics, while various mass spectrometers are used for proteomics and metabolomics. With the spread and advancement of next-generation sequencing technologies, more high-throughput omics analyses can be performed at a lower cost, and statistical tools, such as machine learning, are becoming more widely used, making it possible to integrate multiple omics analyses.

In the case of solid tumors, it is necessary to collect tumor tissue itself by some method for omics analysis. Recently, however, a method called liquid biopsy has sometimes been used to extract the genome from cancer cell-derived DNA (cell-free DNA) [7–9] or circulating tumor cells [10, 11] in the blood for analysis, rather than extracting the cancer genome from the tumor tissue itself. With liquid biopsy, cancer genome information can be obtained only by blood sampling, even in cancer types for which tumor tissue is difficult to obtain, and changes over the course of treatment can also be analyzed because the test can be performed many times [12, 13]. Tissues collected by biopsy or surgery contain not only tumor cells but also stromal cells, lymphocytes, vascular endothelial cells, and many other types of cells, which sometimes interfere with accurate analysis by constituting noise in genome analyses [14–16]. Therefore, a method called single-cell analysis has been developed, in which the collected tumor tissue is separated into single cells, and genomics analysis is performed on each cell [17, 18]. This method makes it possible to analyze the genomic data of each cell, and the characteristics of tumor cells and the relationships between cells are being clarified [19, 20] (Fig. 8.1).

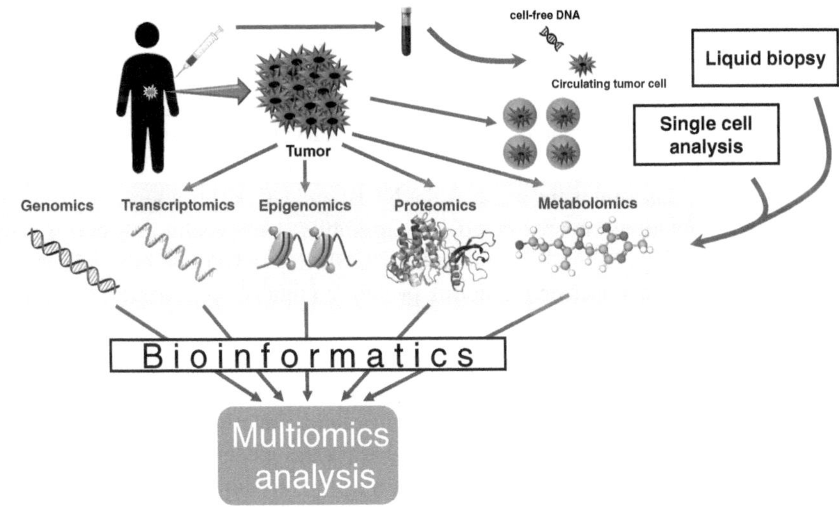


**Fig. 8.1** Concept of omics analysis. Omics analysis is the analysis of the genome, transcriptome, epigenome, proteome, metabolome, and other information obtained from tumors. Bioinformatics is necessary for multiomics analysis that integrates these data. Liquid biopsy, which obtains omics information from circulating tumor cells in the blood, and single-cell analysis, which analyzes single cells constituting a tumor, have been developed, and their integrated analysis is also underway

## 8.3 Omics Databases and Analysis Tools

With the development of next-generation sequencing technology, sequencing has become faster and less expensive, and an enormous number of omics analyses are being conducted worldwide, resulting in an explosive increase in the amount of data accumulated. Conversely, the increase in the number of public databases around the world has made it difficult for users to obtain the data that they need for their research purposes. In Japan, the National Bioscience Database Center (NBDC) was established in 2011 to integrate various life science databases and promote data sharing and utilization. The database catalog [21] is available to the public, facilitating database searches. The main databases are listed below.

- Nucleotides
  GenBank [22, 23] is a nucleic acid sequence database maintained by the National Center for Biotechnology Information (NCBI). It is part of the International Nucleotide Sequence Database Collaboration (INSDC), which is operated by the NCBI, the Nucleotide Archive (ENA) [24], and the DNA Data Bank of Japan (DDBJ) [25], and data are exchanged between these organizations.
- Genomes
  The UCSC Genome Browser [26, 27] is a project of UCSC that automatically annotates eukaryotic organisms with genomes that have been decoded and publishes the results in a database. The genome information used is the same as that of NCBI and Ensembl, but the annotated information is diverse, including originally calculated information and information from NCBI and Ensembl. One of the characteristics of this system is that the annotated information itself is often newer because of the high-speed automatic annotation.

  Ensembl [28, 29] which is a joint project of the European Molecular Biology Laboratory (EMBL)-European Bioinformatics Institute (EBI) and the Sanger Centre, performs automated annotation of eukaryotic organisms with genomes that have been decoded and publishes the results in a database. The information provided by Ensembl is the same as the NCBI and UCSC browsers for genomes, but the annotations are predicted by Ensembl's own pipeline. Therefore, the information differs slightly from that of NCBI Mapviewer and others. The prediction pipeline focuses on predicting protein-coding genes as accurately as possible, so the prediction accuracy is high.

  NCBI Genome [30] is a database of genome information managed and operated by NCBI. In recent years, genome information about many new species of organisms has been registered, and one can quickly determine how much nucleotide sequence information has been revealed for the species in which one is interested. The Genome Data Viewer allows users to visualize molecular data in a genomic context and graphically display data about a given experiment or sample. Genome information about species commonly used in research can be organized for easy visual and understandable retrieval from a phylogenetic tree, or genomes can be compared [31].

- Epigenomes
The International Human Epigenome Consortium (IHEC) is an international consortium that aims to map the human epigenome in relation to various diseases and life phenomena, and the IHEC-Data Portal [32] is populated with data from various databases. By selecting the species, tissue, assay method, and provider, one can view the available datasets in a grid view, track them in the UCSC Genome Browser, and download the data in batches [33].
The NIH Roadmap Epigenomics Mapping Consortium aims to provide epigenomic maps of histone modifications and DNA methylation in various tissues and cell types related to human diseases. The Roadmap Epigenomics Project [34] allows users to browse data by adult, fetal, brain, stem cell, etc., and to view genomic information in the USCS Genome Browser. Protocols, tools, and project information are also provided [35].
- Gene Expression
The NCBI Gene Expression Omnibus (GEO) [36, 37] is a database of gene expression information provided and maintained by NCBI. GEO mainly contains data obtained by microarrays, and the amount of registered data is very large. Not only can one search for gene expression datasets and gene profiles of interest among them, but one can also freely download the raw data.
ArrayExpress [38, 39] is a database of gene expression information provided and maintained by EBI, and similar to NCBI-GEO, it mainly stores data obtained by microarrays and allows users to search for expression datasets and gene profiles and obtain raw data from them.
- Proteins
The Universal Protein Resource (UniProt) [40, 41] operated and maintained by EMBL-EBI, the Swiss Institute of Bioinformatics (SIB), and the Protein Information Resource (PIR), is a database of protein UniProt consisting of UniProt Knowledgebase (UniProtKB), UniProt Reference Clusters (UniRef), and UniProt Archive (UniPrac). UniProtKB publishes SwissProt, which is manually annotated with high-quality annotations based on information from the literature, and TrEMBL (Translated EMBL Nucleotide Sequence Data Library), which is mechanically annotated. UniRef provides the results of preformed sequence homology searches, and UniPrac compiles information, such as IDs of other databases by sequence ID.
InterPro [42, 43] is an integrated database that collects descriptions of protein family classifications, domains, and functional sites based on EBI. It brings together multiple databases that contain the characteristics of various proteins and provide protein characteristics at various levels. Using InterProScan, a database search tool, a single amino acid sequence can be searched in multiple databases integrated by InterPro to efficiently infer protein families and domain repeat structures that match the queried sequence.
The PRIDE (Proteomics Identifications Archive) database [44, 45] is a public repository of proteomics data operated by EMBL-EBI.

- Other
  ENCODE [46] is a database that aims to compile a comprehensive list of functional factor parts of the human genome. It contains information about factors that function at the protein and RNA levels, as well as regulatory factors that control the environment in which cells and genes are activated, including methods of analysis, sample outlines, replicon types, experimental conditions, and analysis flow for the studies [47].

The Cancer Genome Atlas (TCGA) is the largest and most comprehensive cancer genome database launched by the National Institutes of Health (NIH) in 2006. The TCGA dataset contains more than 10,000 cases of 33 different cancer types, and omics information, such as cancer genomes, epigenomes, and transcriptomes, is publicly available. The level at which this omics information is used (raw data or data after being processed by multiple software) depends on the researcher's intended use, but software for multiomics analysis is needed to analyze the data in an integrated manner. Such software is developed by bioinformaticians around the world, and most of it is available as freeware, so it can be installed and used according to the purpose of use, but to use such software, some knowledge of programming languages such as R, Python, Perl, etc., is needed. Many cancer researchers have not mastered the art of bioinformatics analysis and find it difficult to explore TCGA data resources, but several web tools have been developed that allow them to analyze TCGA data. cBioPortal for Cancer Genomics [48–50] is a web tool developed at Memorial Sloan Kettering Cancer Center that integrates omics data from multiple public databases, including TCGA data, and integrates omics and clinical data for analysis and visualization. Broad GDAC Firehose [51] is a pipeline for processing and analyzing large datasets via dozens of quantitative algorithms developed at the Broad Institute and the results of these analyses, which can be explored and visualized using FireBrowse [52]. UCSC Xena [53, 54] is a web browser-based visualization and analysis tool for large public cancer genome datasets from TCGA, ICGC (International Cancer Genome Consortium), GDC (Genomic Data Commons), and other databases. It is possible to freely combine and analyze SNVs (single nucleotide variants), INDELs, large-scale structural variations, CNVs (copy number variations), gene expression, DNA methylation, ATAC-seq, and other data from each database. LinkedOmic [55, 56] includes multiomics data from 32 TCGA carcinomas, as well as proteomics data from breast, ovarian, and colorectal cancer TCGA was generated by the Clinical Proteomics Tumor Analysis Consortium (CPTAC). It is a web tool that analyzes, compares, and makes biological sense of data using three modules: LinkFinder, LinkInterpreter, and LinkCompare. In addition to the above, there are many web tools, including The Cancer Proteome Atlas Portal (TCPA) [57, 58], which is an integrated data portal for analyzing and visualizing TCGA proteomic data; MEXPRESS [59, 60], which enables visualization of TCGA clinical data, gene expression data, and DNA methylation data; and GEPIA2 [61, 62], which can analyze and visualize data from the GTEx (Genotype Tissue Expression) project, which examined gene expression in TCGA, human body tissue, and genotypes. As described, there are a vast number of analysis tools in existence, and it is difficult to determine which tool to use. For major tools, there are often videos on how to use them and the features of the tools, which can be used as a reference.

## 8.4 Cancer Clinical Trials Using Omics Data

With technological advances in omics analysis, molecular markers that are useful for predicting therapeutic efficacy have been identified in various types of cancer. Molecularly targeted drugs that are expected to be effective against patients with such biomarkers have been developed, and their efficacy has been reported in multiple clinical trials. Representative examples include vemurafenib for metastatic malignant melanoma with the BRAF V600E mutation [63], gefitinib for *EGFR* mutation-positive non-small cell lung cancer [64], cetuximab [65], and panitumumab [66] for *KRAS* wild-type colon cancer, crizotinib [67], alectinib [68], and ceritinib [67] for *ALK* fusion gene-positive non-small cell lung cancer, and so on. In the development of such molecular-targeted drugs, which are expected to be effective against a specific biomarker, problems in terms of development cost and time have been considered, such as the need to conduct as many clinical trials as the number of drugs to be developed to verify their efficacy and the need to verify efficacy for each cancer type when biomarkers are detected across multiple cancer types. Therefore, a comprehensive clinical trial protocol called a master protocol has been proposed and implemented in recent years [69]. The draft guidance published by the US FDA in 2018 defined a master protocol as a single protocol designed with multiple subtrials that evaluate the effects of one or more investigational drugs on one or more disease subtypes with different objectives within the framework of the overall study structure and within the overall clinical trial framework. Each subtrial is often categorized by population based on cancer type, histology, and biomarkers, and by conducting each subtrial in parallel based on a comprehensive protocol, more hypotheses can be tested efficiently and in less time.

Master protocols are classified into three categories according to the characteristics of the target population and the type and number of study treatments: basket trials, umbrella trials, and platform trials. Basket trials are trials designed to validate a single investigational drug or drug combination in different populations defined by specific genetic/molecular biomarkers, rather than patient eligibility being limited to a specific cancer type. Thus, each subtrial (basket) is composed of different types of cancer, and each subtrial tests a different treatment (Fig. 8.2). The advantages of a basket trial include the potential to offer patients with a broad range of cancer types a treatment option with a molecular-targeted agent that might not have been tested in clinical trials for their disease, the short time from initial diagnosis and eligibility to subsequent cohort assignment and initiation of treatment, and the

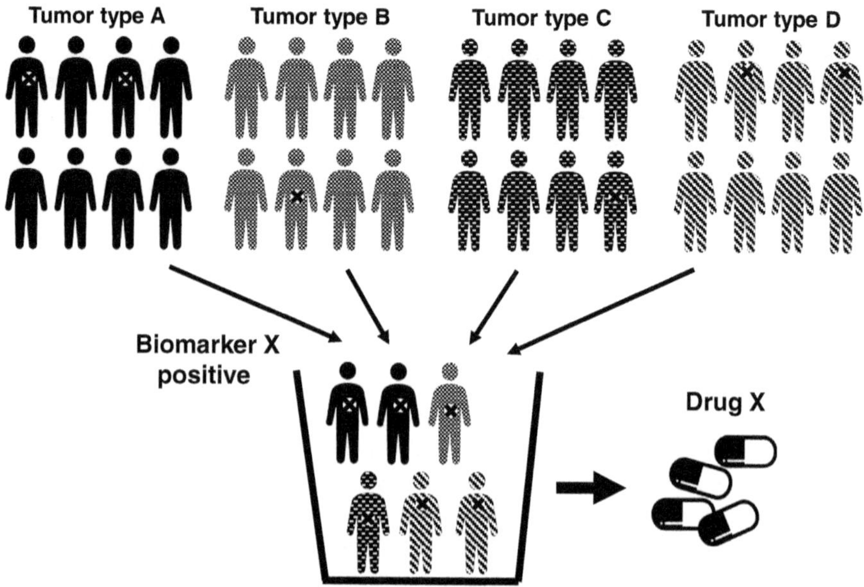


**Fig. 8.2** Scheme of basket trials. Basket trials are master protocols for targeted therapy based on specific biomarkers from multiple cancer types. Each subtrial is often a single-arm exploratory study

often small number of patients in each cohort, resulting in a short time to results being reported. One problem is that the basket trial assumes that classification by molecular characteristics of the tumor can substitute for classification by tumor histology but that histology might be a stronger predictor of response to targeted therapy than biomarkers [70].

Umbrella trials evaluate targeted therapies in specific cancer types by assigning patients to one of a number of subtrials defined by genetic mutations or biomarkers. Subtrials are often single-arm or randomized subtrials for validation purposes, whereas basket trials are generally single-arm subtrials for exploratory purposes (Fig. 8.3). By fixing the cancer type of interest, umbrella trials are able to draw cancer-specific conclusions with less heterogeneity that might exist within a given cohort compared to basket trials. In addition, randomized trials of targeted and non-targeted therapies in subtrials can evaluate the presumed mechanism of action of a therapeutic agent and empirically distinguish between prognostic and efficacy-predicting markers. However, the feasibility of targeting a single cancer type creates problems. Particularly for rare diseases, allocation to subtrials by biomarker can slow enrollment within a cohort and thus slow trial progression. There is also the challenge that, if a large, long-term protocol design is needed, changes in treatment status, such as the emergence of a new standard of care during the period, might render the subtrial less clinically meaningful in its original setting, further lengthening the duration of the trial [70].

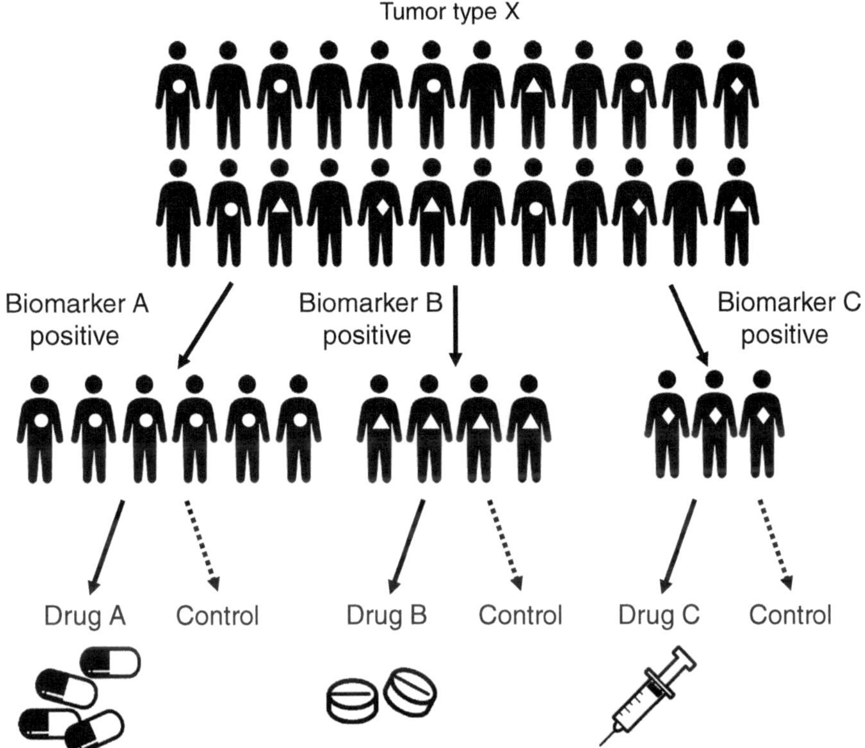


**Fig. 8.3** Scheme of umbrella trials. Umbrella trials are master protocols in which targeted therapy is administered in each subtest defined by multiple biomarkers for a specific cancer type. Each subtrial is a single-arm or randomized trial and is often a validation trial

Platform trials are a generic term for randomized designs with a common control group and several different targeted treatment groups and trials that allow for the addition or exclusion of new treatments or eligible patients during the trial (Fig. 8.4). Platform trials evaluate the efficacy and futility of each targeted therapy in an interim analysis, and the treatment effect is often modeled as an independent parameter across biomarker-defined subtypes according to a Bayesian hierarchical model. Platform trials are often long-term trials because new trials can be added, and as with umbrella trials, the standard of care can change during the trial period due to the emergence of new treatments (and possibly the trial itself, which was originally conducted) [71]. In such cases, the protocol, statistical analysis plan, informed consent document, etc., might need to be modified, and the trial might have to be suspended [72].

Examples of these master protocol trials are listed below:

- Basket Trials

  The NCI-MATCH (NCI Molecular Analysis for Therapeutic Choice) trial consists of 24 substudies evaluating the efficacy of at least 17 targeted therapies in patients with solid tumors and lymphomas who have received at least one regi-

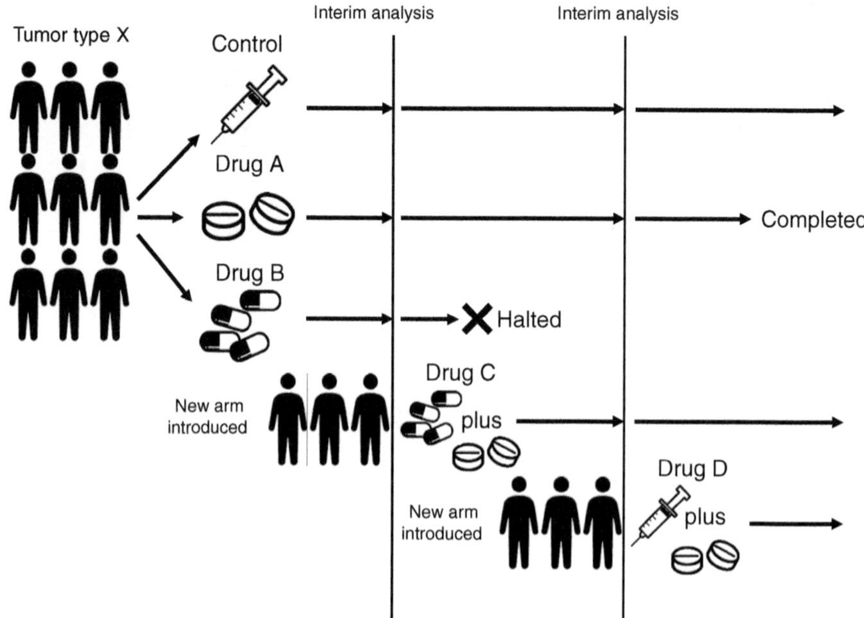


**Fig. 8.4** Scheme of platform trials. Platform trials are randomized designs, in which several different targeted therapy arms share a common control group. New target treatment groups can be added during the trial, and existing treatment groups can be excluded

men of therapy. The primary endpoint of each subtrial was tumor shrinkage, with a single-arm design based on a binomial distribution to enroll 35 patients and an important secondary endpoint of 6-month progression-free survival [73]. The NCI-MATCH study can be interpreted as a basket study because it evaluates the efficacy and safety of targeted agents across cancer types in a molecular marker-positive population. In contrast, subtrials might be conducted to evaluate the efficacy and safety of multiple targeted agents against a molecular marker of interest in a specific cancer type, which can be interpreted as an umbrella study. Thus, the NCI-MATCH study can be described as a study with the characteristics of both a basket study and an umbrella study.

The AcSé study is a phase II study of various solid tumors (e.g., gastrointestinal, breast, kidney, ovarian, and thyroid cancers), consisting of 23 substudies evaluating the efficacy and safety of crizotinib alone in patients with at least one *ALK*, *MET*, *RON*, or *ROS-1* mutation [74]. Each substudy is defined by mutation and pathology and is designed according to a two-stage design. NSCLC with *ROS-1* translocation and esophageal/gastric cancer with *MET* amplification has been reported thus far [75, 76].

The KEYNOTE-158 trial is a phase II study of solid tumors refractory to standard chemotherapy with MSI-high—a condition in which microsatellite instability (MSI) due to abnormal DNA mismatch repair (dMMR) is frequently observed, except for unresectable or metastatic colorectal cancer—which evaluated the efficacy and safety of pembrolizumab, an anti-PD-1 antibody. The primary endpoint was the objective response rate (ORR), with a median ORR of 34.3% [77].

- Umbrella Studies
  The ALCHEMIST (The Adjuvant Lung Cancer Enrichment Marker Identification and Sequencing Trial) trial is a randomized umbrella trial for patients with *ALK*- or *EGFR*-positive high-risk lung adenocarcinoma. Patients with *ALK*- or *EGFR*-positive disease will be enrolled in a randomized phase III subtrial of crizotinib versus placebo or erlotinib versus placebo. The primary endpoint of each trial will be overall survival, and interim analyses are planned; if both *ALK* and *EGFR* are negative, PD-L1 expression will be measured, and enrollment in a randomized subtrial of nivolumab plus observation will be considered. The primary endpoints of this subtrial are overall survival and disease-free survival [78].

  The Lung-MAP trial was initiated as an umbrella study to test the efficacy of multiple targeted therapies in advanced or recurrent squamous non-small cell lung cancer. All subtrials were designed as randomized phase II/III or single-arm phase II trials, with biomarker screening resulting in taselisib being assigned for *PIK3CA*-positive patients, palbociclib for patients positive for cell cycle gene mutations, rilotumumab plus erlotinib for patients positive for *c-MET,* and ADZ4547 for *FGFR*-positive patients; patients with positive homologous recombination repair abnormalities were assigned to tarazoparib and a control group. Biomarker-negative patients were randomized to durvalumab plus docetaxel, nivolumab plus ipilimumab or nivolumab alone for anti-PD-(L)1 therapy-naive patients as an unmatched subtrial and durvalumab plus tremelimumab for patients relapsing after anti-PD-(L)1 therapy. The primary endpoint of the substudy was progression-free survival or overall survival [79].

  The plasmaMATCH trial is a nonrandomized, phase IIa trial to test the efficacy of targeted therapy in advanced recurrent breast cancer by detecting targeted gene mutations and testing circulating tumor DNA (ctDNA). ctDNA testing identified *ESR1*, *HER2*, *AKT1*, and *PTEN* mutations. Patients were classified into four cohorts according to mutations and tumor estrogen receptor status and were treated with fulvestrant, neratinib, and capibasertib as single agents or in combination. The primary endpoint was the objective response rate [80].
- Platform Trials
  The FOCUS4 trial is a placebo-controlled, multiarm, multistage, randomized trial testing the efficacy of multiple targeted therapies for untreated colorectal cancer. In a population of patients with specific molecular markers, safety is evaluated in the first stage, proof of concept is confirmed in the second stage, short-term efficacy is evaluated in the third stage, and long-term efficacy is evaluated in the fourth stage. The efficacy endpoints are progression-free survival and overall survival. In such a multiarm, multistage trial, new treatments can be added during the trial, or treatments that prove to be futile can be excluded before the third or fourth stage, corresponding to a Phase III trial. All FOCUS4 trials are open to patients with negative molecular markers [81].

The STAMPEDE trial is a randomized platform trial with a multiarm, multistage design in high-risk prostate cancer patients. The trial was originally initiated as a five-arm study comparing a control group with a single agent or a combination of zoledronic acid, docetaxel, and celecoxib in patients initiating hormone therapy [82]. It was subsequently modified multiple times, with the addition of treatment groups with abiraterone and enzalutamide administered as single agents or in combination with radiation therapy [83, 84]. The current protocol continues to study metformin and transdermal estradiol.

The I-SPY2 trial is a phase II, adaptive, randomized, controlled trial evaluating the efficacy of a new investigational agent in combination with standard neoadjuvant chemotherapy for stage II/III high-risk breast cancer [85]. In this trial, enrolled patients will be classified into ten molecular subtypes and assigned to a study arm according to subtype based on an adaptive randomization engine using predictive probability. The primary endpoint is pathologic complete response (pCR) or a residual cancer burden (RCB) of 0. A Bayesian design is used, in which the predictive probabilities are updated as needed based on treatment results, and new predictive probabilities are assigned. When the predicted probability of a study drug reaches a predefined level of efficacy in one or more subtypes, the drug is "graduated" and proceeds to Phase III trials. Up to five study drugs can be evaluated in parallel at the same time, including combinations. To date, graduation has occurred for neratinib [86], veliparib with carboplatin [87], MK-2206 [88], and pembrolizumab [89].

There are several other issues that have been discussed regarding cancer clinical trials with master protocols, in addition to those listed in the brief description of each trial. Ethical issues include that the complexity and duration of the trials cause the informed consent documents to be more complex, so patients might not be able to understand the documents or the trial concept itself to the degree necessary to provide correct informed consent. In addition, in trials in which the concept of a final dose has not yet been fully established, the adaptive plan could lead to the abandonment of suboptimal dose regimens as the trial progresses, and the benefit-risk ratio might change during the dose optimization process [72]. A master protocol is a single clinical trial that encompasses multiple subtrials, but each subtrial is independently validated, requiring very sophisticated and complex statistics. For example, statistical power could be lost if a single master protocol is hypothesized to be accepted by the results of the subtrials, resulting in the closure or opening of study groups even though no adjustments are specified in the protocol. Another complication noted for the control of Type I error is that multiplicity adjustment might be required for one treatment comparison but not for another [71, 90]. However, master protocols offer tremendous advantages in flexibility and efficiency in drug development, and it is anticipated that many trials will be designed in personalized therapy using cancer genomics data. Trial designs are also expected to increase the sensitive allocation of patients to matched therapies, including combination therapies according to multiple driver mutations, biomarkers, and pathways. Master protocols could also provide insights into the molecular mechanisms of exceptional responders, in whom drugs that are not effective in other patients are found to be significantly effective, which could be useful in designing future master protocols for specific disease types.

## 8.5 Genomic Analysis and Personalized Treatment in Gynecologic Cancer

There are four types of gynecological cancers for which integrated genome analysis was performed by TCGA—ovarian cancer (high-grade serous carcinoma; HGSC), endometrial cancer, cervical cancer, and uterine sarcoma—and analysis results have been reported for HGSC [91], endometrial cancer [92], and cervical cancer [93].

- HGSC
  Whole-exome sequencing analysis of 316 HGSCs identified *TP53* somatic mutations in 96% of HGSCs. In addition, mutations in the *BRCA1/2* gene were found in approximately 20% of both germline and somatic cases. BRCA1/2 is involved in the DNA homologous recombination repair pathway, and mutations in genes encoding proteins involved in homologous recombination repair other than BRCA1/2 were also observed in HGSCs, suggesting that approximately 50% of HGSCs have abnormal homologous recombination repair (HRD).
- Endometrial Cancer
  Genomic analysis of 373 cases of endometrial cancer classified cancers into the following four categories: (1) POLE type (ultramutated) with a very high frequency of gene mutations; (2) MSI type (hypermutated) with a high frequency of gene mutations and methylation of the *MLH1* promoter region in many of them; (3) copy number low type (endometrioid), in which the frequency of mutations is low and microsatellite stable; and (4) copy number high type (serous-like), consisting mainly of serous-like tumors with significant copy number changes and low frequency of genetic mutations.
- Cervical Cancer
  Integrated genomic analysis of 228 cervical cancer cases revealed genomic alterations in either or both the PI3K-MAPK and TGFβ signaling pathways in more than 70% of cases. In addition, amplification of the *CD274* gene encoding PD-L1 and the *PDCD1LG2* gene encoding PD-L2 was observed in approximately 20% of the cases.

Although the integrated genome analysis of these three gynecological cancers has revealed new cancer genome features, there are only two targeted therapies in practical use in the field of gynecological cancer: PARP inhibitors (olaparib, niraparib, etc.) for *BRCA1/2* mutations or HRD-positive ovarian cancer [94–96]; and PD-1 inhibitors (pembrolizumab) for endometrial cancer with dMMR [97]. Compared to other types of cancer, personalized treatment in gynecological cancer has not progressed very much. However, efficient clinical trial designs for new therapies based on master protocols have recently become possible, and new clinical trials based on master protocols are being conducted in gynecological cancers.

The AMBITION trial is an umbrella study of platinum-resistant recurrent ovarian cancer that uses HRD and PD-L1 biomarkers to allocate treatment groups, with HRD-positive patients receiving olaparib plus cediranib or durvalumab and HRD-negative patients receiving durvalumab plus chemotherapy or durvalumab plus tremelimumab plus chemotherapy based on PD-L1 expression; the primary endpoint is the objective response rate [98]. The BOUQUET trial (NCT04931342) [99] is a biomarker-driven phase II study of recurrent epithelial ovarian cancer in patients with non-high-grade serous carcinoma and non-high-grade endometrial carcinoma, including ipatasertib plus paclitaxel for patients with *PIK3CA/AKT1/PTEN* mutations, obimetinib for patients with *BRAF/NRAS/KRAS/NF-1* mutations, trastuzumab emtansine for patients with *ERBB2* amplification or mutations, and atezolizumab plus bevacizumab for the unmatched group. The primary endpoint is the objective response rate, and the study is designed as a platform trial. In addition, a project to develop a new adaptive platform trial called Ovarian CanceRx was announced in 2021 [100], and master protocol trials for recurrent ovarian cancer are expected to increase in the future. In endometrial cancer, a phase II umbrella study is underway of retifanlimab alone or in combination with epacadostat or pemigatinib in patients with advanced or metastatic endometrial cancer that has progressed on or after platinum-based chemotherapy (NCT04463771) [101], and more clinical trials based on similar master protocols are expected to follow.

## 8.6 Conclusion

Cancer genomics (omics) analysis is expected to become more comprehensive and detailed in the future, and more personalized medicine based on cancer characteristics is being sought. As cancer research progresses, a new clinical trial framework called "master protocols" has been proposed and implemented to promote more efficient and flexible clinical trials. With the rapid evolution of cancer omics analysis from bulk tumor analysis to single-cell analysis and from single-omics analysis to multiomics analysis, it is highly likely that such omics analysis will be applied in clinical practice. Although it is impossible for clinicians to examine each individual datum, it is necessary to accumulate knowledge to prepare for the advent of omics medicine in the trend of genomic medicine.

## References

1. Winkler H. Verbreitung und Ursache der Parthenogenesis im Pflanzen und Tierreiche. Jena: Fisher; 1920.
2. Kihara H, Lilienfelt F. Genomanalyse bei *Triticum* und *Aegilops*. IV. Untersuchungen as *Aegilops* x *Triticum*- und *Aegilops* x *Aegilops*-Bastarden. Cytologia (Tokyo). 1932;3:384–456.

3. de Anda-Jáuregui G, Hernández-Lemus E. Computational oncology in the multi-omics era: state of the art. Front Oncol. 2020;10:423.
4. Hanahan D, Weinberg RA. Hallmarks of cancer: the next generation. Cell. 2011;144:646–74.
5. Chakraborty S, Hosen MI, Ahmed M, Shekhar HU. Onco-multi-OMICS approach: a new frontier in cancer research. Biomed Res Int. 2018;2018:1–14.
6. Hasin Y, Seldin M, Lusis A. Multi-omics approaches to disease. Genome Biol. 2017;18:83.
7. Jahr S, Hentze H, Englisch S, Hardt D, Fackelmayer FO, Hesch RD, Knippers R. DNA fragments in the blood plasma of cancer patients: quantitations and evidence for their origin from apoptotic and necrotic cells. Cancer Res. 2001;61:1659–65.
8. Schwarzenbach H, Hoon DSB, Pantel K. Cell-free nucleic acids as biomarkers in cancer patients. Nat Rev Cancer. 2011;116(11):426–37.
9. Alix-Panabières C, Pantel K. Liquid biopsy: from discovery to clinical applicationliquid biopsy: from discovery to clinical application. Cancer Discov. 2021;11:858–73.
10. Cristofanilli M, Budd GT, Ellis MJ, et al. Circulating tumor cells, disease progression, and survival in metastatic breast cancer. N Engl J Med. 2009;351:781–91. https://doi.org/10.1056/NEJMoa040766.
11. Pantel K, Alix-Panabières C. Circulating tumour cells in cancer patients: challenges and perspectives. Trends Mol Med. 2010;16:398–406.
12. Haber DA, Velculescu VE. Blood-based analyses of cancer: circulating tumor cells and circulating tumor DNA. Cancer Discov. 2014;4:650–61.
13. Sparano J, O'Neill A, Alpaugh K, Wolff AC, Northfelt DW, Dang CT, Sledge GW, Miller KD. Association of circulating tumor cells with late recurrence of estrogen receptor–positive breast cancer: a secondary analysis of a randomized clinical trial. JAMA Oncol. 2018;4:1700–6.
14. Heng HHQ, Stevens JB, Bremer SW, Liu G, Abdallah BY, Ye CJ. Evolutionary mechanisms and diversity in cancer. Adv Cancer Res. 2011;112:217–53.
15. Lee JH, Daugharthy ER, Scheiman J, et al. Highly multiplexed subcellular RNA sequencing in situ. Science. 2014;343:1360–3.
16. Ahmed R, Zaman T, Chowdhury F, Mraiche F, Tariq M, Ahmad IS, Hasan A. Single-cell RNA sequencing with spatial transcriptomics of cancer tissues. Int J Mol Sci. 2022;23:3042.
17. Shalek AK, Satija R, Adiconis X, et al. Single-cell transcriptomics reveals bimodality in expression and splicing in immune cells. Nature. 2013;498:236–40. https://doi.org/10.1038/NATURE12172.
18. Patel AP, Tirosh I, Trombetta JJ, et al. Single-cell RNA-seq highlights intratumoral heterogeneity in primary glioblastoma. Science. 2014;344:1396–401.
19. Azizi E, Carr AJ, Plitas G, et al. Single-cell map of diverse immune phenotypes in the breast tumor microenvironment. Cell. 2018;174:1293–1308.e36.
20. Lambrechts D, Wauters E, Boeckx B, et al. Phenotype molding of stromal cells in the lung tumor microenvironment. Nat Med. 2018;248(24):1277–89.
21. Home. Integbio database catalog. https://integbio.jp/dbcatalog/?lang=en. Accessed 31 Mar 2022.
22. Sayers EW, Cavanaugh M, Clark K, Ostell J, Pruitt KD, Karsch-Mizrachi I. GenBank. Nucleic Acids Res. 2020;48:D84–6.
23. GenBank overview. https://www.ncbi.nlm.nih.gov/genbank/. Accessed 31 Mar 2022.
24. Harrison PW, Ahamed A, Aslam R, et al. The European nucleotide archive in 2020. Nucleic Acids Res. 2021;49:D82–5.
25. Okido T, Kodama Y, Mashima J, Kosuge T, Fujisawa T, Ogasawara O. DNA Data Bank of Japan (DDBJ) update report 2021. Nucleic Acids Res. 2022;50:D102–5.
26. Gonzalez JN, Zweig AS, Speir ML, et al. The UCSC Genome Browser database: 2021 update. Nucleic Acids Res. 2021;49:D1046–57.
27. UCSC Genome browser home. https://genome.ucsc.edu/. Accessed 31 Mar 2022.
28. Howe KL, Achuthan P, Allen J, et al. Ensembl 2021. Nucleic Acids Res. 2021;49:D884–91.
29. Ensembl genome browser 105. https://www.ensembl.org/index.html. Accessed 31 Mar 2022.

30. Home. Genome. NCBI. https://www.ncbi.nlm.nih.gov/genome/. Accessed 31 Mar 2022.
31. Rangwala SH, Kuznetsov A, Ananiev V, et al. Accessing NCBI data using the NCBI Sequence Viewer and Genome Data Viewer (GDV). Genome Res. 2021;31:159–69.
32. Data Portal. IHEC. https://epigenomesportal.ca/ihec/. Accessed 31 Mar 2022.
33. Bujold D, de Lima Morais DA, Gauthier C, et al. The international human epigenome consortium data portal. Cell Syst. 2016;3:496–499.e2.
34. Roadmap Epigenomics Project. Home. http://www.roadmapepigenomics.org/. Accessed 31 Mar 2022.
35. Bernstein BE, Stamatoyannopoulos JA, Costello JF, et al. The NIH roadmap epigenomics mapping consortium. Nat Biotechnol. 2010;28:1045–8.
36. Home. GEO. NCBI. https://www.ncbi.nlm.nih.gov/geo/. Accessed 31 Mar 2022.
37. Barrett T, Wilhite SE, Ledoux P, et al. NCBI GEO: archive for functional genomics data sets—update. Nucleic Acids Res. 2013; https://doi.org/10.1093/NAR/GKS1193.
38. ArrayExpress < EMBL-EBI. https://www.ebi.ac.uk/arrayexpress/. Accessed 31 Mar 2022.
39. Athar A, Füllgrabe A, George N, et al. ArrayExpress update—from bulk to single-cell expression data. Nucleic Acids Res. 2019;47:D711–5.
40. UniProt. https://www.uniprot.org/. Accessed 31 Mar 2022.
41. Bateman A, Martin MJ, Orchard S, et al. UniProt: the universal protein knowledgebase in 2021. Nucleic Acids Res. 2021;49:D480–9.
42. InterPro. https://www.ebi.ac.uk/interpro/. Accessed 31 Mar 2022.
43. Marabelle A, Le DT, Ascierto PA, et al. The InterPro protein families and domains database: 20 years on. Nucleic Acids Res. 2021;49:1274–84.
44. PRIDE. Proteomics identification database. https://www.ebi.ac.uk/pride/. Accessed 31 Mar 2022.
45. Perez-Riverol Y, Bai J, Bandla C, et al. The PRIDE database resources in 2022: a hub for mass spectrometry-based proteomics evidences. Nucleic Acids Res. 2022;50:D543–52.
46. ENCODE. https://www.encodeproject.org/. Accessed 31 Mar 2022.
47. Abascal F, Acosta R, Addleman NJ, et al. Perspectives on ENCODE. Nature. 2020;583:693–8.
48. Cerami E, Gao J, Dogrusoz U, et al. The cBio cancer genomics portal: an open platform for exploring multidimensional cancer genomics data: figure 1. Cancer Discov. 2012;2:401–4.
49. Gao J, Aksoy BA, Dogrusoz U, et al. Integrative analysis of complex cancer genomics and clinical profiles using the cBioPortal. Sci Signal. 2013;6:pl1.
50. cBioPortal for Cancer Genomics. https://www.cbioportal.org/. Accessed 31 Mar 2022.
51. Broad GDAC Firehose. https://gdac.broadinstitute.org/. Accessed 31 Mar 2022.
52. FireBrowse. http://firebrowse.org/. Accessed 31 Mar 2022.
53. Goldman MJ, Craft B, Hastie M, et al. Visualizing and interpreting cancer genomics data via the Xena platform. Nat Biotechnol. 2020;38:1–4.
54. UCSC Xena. http://xena.ucsc.edu/. Accessed 31 Mar 2022.
55. LinkedOmics. Login. http://www.linkedomics.org/login.php. Accessed 31 Mar 2022.
56. Vasaikar SV, Straub P, Wang J, Zhang B. LinkedOmics: analyzing multi-omics data within and across 32 cancer types. Nucleic Acids Res. 2018; https://doi.org/10.1093/NAR/GKX1090.
57. TCPA. Home. https://tcpaportal.org/tcpa/. Accessed 31 Mar 2022.
58. Li J, Lu Y, Akbani R, et al. TCPA: a resource for cancer functional proteomics data. Nat Methods. 2013;10:1046–7.
59. MEXPRESS. https://mexpress.be/. Accessed 31 Mar 2022.
60. Koch A, Jeschke J, Van Criekinge W, van Engeland M, De Meyer T. MEXPRESS update 2019. Nucleic Acids Res. 2019;47:W561–5.
61. GEPIA 2. http://gepia2.cancer-pku.cn/#index. Accessed 31 Mar 2022.
62. Tang Z, Kang B, Li C, Chen T, Zhang Z. GEPIA2: an enhanced web server for large-scale expression profiling and interactive analysis. Nucleic Acids Res. 2019;47:W556–60.
63. Chapman PB, Hauschild A, Robert C, et al. Improved survival with vemurafenib in melanoma with BRAF V600E mutation. N Engl J Med. 2011;364:2507–16.
64. Mok TS, Wu Y-L, Thongprasert S, et al. Gefitinib or carboplatin–paclitaxel in pulmonary adenocarcinoma. N Engl J Med. 2009;361:947–57.

65. Jonker DJ, O'Callaghan CJ, Karapetis CS, et al. Cetuximab for the treatment of colorectal cancer. N Engl J Med. 2007;357:2040–8.
66. Amado RG, Wolf M, Peeters M, et al. Wild-type *KRAS* is required for panitumumab efficacy in patients with metastatic colorectal cancer. J Clin Oncol. 2008;26:1626–34.
67. Shaw AT, Kim D-W, Mehra R, et al. Ceritinib in *ALK* -rearranged Non–small-cell lung cancer. N Engl J Med. 2014;370:1189–97.
68. Shaw AT, Gandhi L, Gadgeel S, et al. Alectinib in *ALK*-positive, crizotinib-resistant, non-small-cell lung cancer: a single-group, multicentre, phase 2 trial. Lancet Oncol. 2016;17:234–42.
69. Woodcock J, LaVange LM. Master protocols to study multiple therapies, multiple diseases, or both. N Engl J Med. 2017;377:62–70.
70. Renfro LA, Sargent DJ. Statistical controversies in clinical research: basket trials, umbrella trials, and other master protocols: a review and examples. Ann Oncol. 2017;28:34–43.
71. Lu CC, Li XN, Broglio K, Bycott P, Jiang Q, Li X, McGlothlin A, Tian H, Ye J. Practical considerations and recommendations for master protocol framework: basket, umbrella and platform trials. Ther Innov Regul Sci. 2021;55:1145.
72. Sudhop T, Brun NC, Riedel C, Rosso A, Broich K, Senderovitz T. Master protocols in clinical trials: a universal Swiss Army knife? Lancet Oncol. 2019;20:e336–42.
73. Flaherty KT, Gray RJ, Chen AP, et al. Molecular landscape and actionable alterations in a genomically guided cancer clinical trial: national cancer institute molecular analysis for therapy choice (NCI-MATCH). J Clin Oncol. 2020;38:3883–94.
74. Buzyn A, Blay JY, Hoog-Labouret N, Jimenez M, Nowak F, Le Deley MC, Pérol D, Cailliot C, Raynaud J, Vassal G. Equal access to innovative therapies and precision cancer care. Nat Rev Clin Oncol. 2016;13:385–93.
75. Moro-Sibilot D, Cozic N, Pérol M, et al. Crizotinib in c-MET- or ROS1-positive NSCLC: results of the AcSé phase II trial. Ann Oncol. 2019;30:1985–91.
76. Aparicio T, Cozic N, de la Fouchardière C, et al. The activity of crizotinib in chemo-refractory MET-amplified esophageal and gastric adenocarcinomas: results from the AcSé-Crizotinib program. Target Oncol. 2021;16:381–8.
77. Marabelle A, Le DT, Ascierto PA, et al. Efficacy of pembrolizumab in patients with non-colorectal high microsatellite instability/mismatch repair–deficient cancer: results from the phase II KEYNOTE-158 study. J Clin Oncol. 2020;38:1–10.
78. Govindan R, Mandrekar SJ, Gerber DE, et al. ALCHEMIST trials: a golden opportunity to transform outcomes in early stage non–small cell lung cancer. Clin Cancer Res. 2015;21:5439.
79. Herbst RS, Gandara DR, Hirsch FR, et al. Lung master protocol (Lung-MAP)—a biomarker-driven protocol for accelerating development of therapies for squamous cell lung cancer: SWOG S1400. Clin Cancer Res. 2015;21:1514–24.
80. Turner NC, Kingston B, Kilburn LS, et al. Circulating tumour DNA analysis to direct therapy in advanced breast cancer (plasmaMATCH): a multicentre, multicohort, phase 2a, platform trial. Lancet Oncol. 2020;21:1296–308.
81. Kaplan R, Maughan T, Crook A, Fisher D, Wilson R, Brown L, Parmar M. Evaluating many treatments and biomarkers in oncology: a new design. J Clin Oncol. 2013;31:4562–8.
82. James ND, Sydes MR, Clarke NW, et al. Systemic therapy for advancing or metastatic prostate cancer (STAMPEDE): a multi-arm, multistage randomized controlled trial. BJU Int. 2009;103:464–9.
83. James ND, Sydes MR, Clarke NW, et al. Addition of docetaxel, zoledronic acid, or both to first-line long-term hormone therapy in prostate cancer (STAMPEDE): survival results from an adaptive, multiarm, multistage, platform randomised controlled trial. Lancet. 2016;387:1163–77.
84. James ND, de Bono JS, Spears MR, et al. Abiraterone for prostate cancer not previously treated with hormone therapy. N Engl J Med. 2017;377:338–51.
85. Barker AD, Sigman CC, Kelloff GJ, Hylton NM, Berry DA, Esserman LJ. I-SPY 2: an adaptive breast cancer trial design in the setting of neoadjuvant chemotherapy. Clin Pharmacol Ther. 2009;86:97–100.

86. Park JW, Liu MC, Yee D, et al. Adaptive randomization of neratinib in early breast cancer. N Engl J Med. 2016;375:11–22.
87. Rugo HS, Olopade OI, DeMichele A, et al. Adaptive randomization of veliparib–carboplatin treatment in breast cancer. N Engl J Med. 2016;375:23.
88. Chien AJ, Tripathy D, Albain KS, et al. MK-2206 and standard neoadjuvant chemotherapy improves response in patients with human epidermal growth factor receptor 2-positive and/or hormone receptor-negative breast cancers in the I-SPY 2 trial. J Clin Oncol. 2020;38:1059–69.
89. Nanda R, Liu MC, Yau C, et al. Effect of pembrolizumab plus neoadjuvant chemotherapy on pathologic complete response in women with early-stage breast cancer: an analysis of the ongoing phase 2 adaptively randomized I-SPY2 trial. JAMA Oncol. 2020;6:676–84.
90. Di Liello R, Piccirillo MC, Arenare L, Gargiulo P, Schettino C, Gravina A, Perrone F. Master protocols for precision medicine in oncology: overcoming methodology of randomized clinical trials. Life (Basel, Switzerland); 2021. https://doi.org/10.3390/LIFE11111253.
91. Cancer Genome Atlas Research Network. Integrated genomic analyses of ovarian carcinoma. Nature. 2011;474:609–15.
92. Getz G, Gabriel SB, Cibulskis K, et al. Integrated genomic characterization of endometrial carcinoma. Nature. 2013;4977447(497):67–73.
93. Burk RD, Chen Z, Saller C, et al. Integrated genomic and molecular characterization of cervical cancer. Nature. 2017;5437645(543):378–84.
94. Pujade-Lauraine E, Ledermann JA, Selle F, et al. Olaparib tablets as maintenance therapy in patients with platinum-sensitive, relapsed ovarian cancer and a BRCA1/2 mutation (SOLO2/ENGOT-Ov21): a double-blind, randomised, placebo-controlled, phase 3 trial. Lancet Oncol. 2017;18:1274–84.
95. Moore K, Colombo N, Scambia G, et al. Maintenance olaparib in patients with newly diagnosed advanced ovarian cancer. N Engl J Med. 2018;379:2495–505.
96. González-Martín A, Pothuri B, Vergote I, et al. Niraparib in patients with newly diagnosed advanced ovarian cancer. N Engl J Med. 2019;381:2391–402.
97. O'Malley DM, Bariani GM, Cassier PA, et al. Pembrolizumab in patients with microsatellite instability-high advanced endometrial cancer: results from the KEYNOTE-158 study. J Clin Oncol. 2022;40:752–61.
98. Lee JY, Yi JY, Kim HS, Lim J, Kim S, Nam BH, Kim HS, Kim JW, Choi CH, Kim BG. An umbrella study of biomarker-driven targeted therapy in patients with platinum-resistant recurrent ovarian cancer: a Korean Gynecologic Oncology Group study (KGOG 3045), AMBITION. Jpn J Clin Oncol. 2019;49:789–92.
99. A study evaluating the efficacy and safety of biomarker-driven therapies in patients with persistent or recurrent rare epithelial ovarian tumors. Full text view. ClinicalTrials.gov. https://clinicaltrials.gov/ct2/show/NCT04931342. Accessed 31 Mar 2022.
100. Ovarian CanceRx. Global Coalition for Adaptive Research. https://www.gcaresearch.org/ovariancancerx/. Accessed 31 Mar 2022.
101. Safety and efficacy of retifanlimab (INCMGA00012) alone or in combination with other therapies in participants with advanced or metastatic endometrial cancer who have progressed on or after platinum-based chemotherapy. Full text view. ClinicalTrials.gov. https://clinicaltrials.gov/ct2/show/NCT04463771. Accessed 31 Mar 2022.

# Chapter 9
# Personalized Treatment of Gynecological Cancer According to Age and Symptom Benefit

**Yoshio Yoshida and Daisuke Inoue**

**Abstract** Since established guidelines for the treatment of gynecological cancers in the elderly are lacking, decisions regarding the treatment plan and policy are currently made in consultation with the patient and family based on the doctor's experience, referring to data from non-elderly and healthy elderly patients. Recently, a new concept of individualized treatment for gynecological cancer in the elderly has started to be established. In the treatment of cancer in the elderly, patients who can receive the standard treatment applicable to healthy non-elderly patients are considered "fit," while those who cannot receive such treatment are considered "unfit." Judgment of suitability requires a comprehensive evaluation of not only individual differences and calendar age, but also physical and mental disabilities, social and economic problems, and medical policy decisions in the case of elderly patients. In particular, when cognitive functions are declining, confirmation of the patient's intentions can represent an obstacle to determining a treatment plan. In such cases, a comprehensive evaluation of the elderly is used to achieve functional assessment. The purpose of this review was to clarify the current status of surgery and chemotherapy for elderly gynecological cancer patients and to discuss future issues.

**Keywords** Elderly cancer patients · Comprehensive geriatric assessment · Geriatric assessment

Y. Yoshida (✉)
Department of Obstetrics and Gynecology, University of Fukui, Fukui, Japan
e-mail: yyoshida@u-fukui.ac.jp

D. Inoue
Faculty of Medical Sciences, Department of Obstetrics and Gynecology, University of Fukui, Fukui, Japan



M. Mandai (ed.), *Personalization in Gynecologic Oncology*, Comprehensive Gynecology and Obstetrics, https://doi.org/10.1007/978-981-19-4711-7_9

## 9.1 Introduction

Cancer is recognized as a common disease of the elderly, with more than 50% of new cases diagnosed after 65 years old, and more than 70% of cancer deaths occurring in this age strata [1]. Ovarian cancer is the seventh most common cancer among women worldwide, accounting for approximately 40% of new cancer cases in women [2]. Ovarian cancer has been reported to have an extremely poor prognosis among gynecological cancers [3], and is also the eighth most common cause of cancer deaths worldwide. Approximately 50% of ovarian cancers are diagnosed among women over 65 years old [1, 3]. This percentage is expected to increase in the coming decades as the population ages and life expectancy increases. Results from the EUROCARE-5 study showed that 5-year survival rates for women diagnosed between 2000 and 2007 were 57% overall, 82% for breast cancer, 76% for uterine cancer, 62% for cervical cancer, 38% for ovarian cancer, 40% for vaginal cancer, and 62% for vulval cancer. Survival rates decreased with increasing age and were more pronounced for ovarian cancer (71% for 15–44 years old, 20% for ≥75 years old) and breast cancer (86% and 72%, respectively) [3].

Various theories have been proposed to explain the lower survival rates of elderly women with cancer, including (a) the older the patient, the more aggressive cancer, including higher grade and more advanced stage; (b) cancers occurring in elderly women are more resistant to chemotherapy; (c) multiple medical comorbidities and individual patient factors such as polypharmacy, functional dependence, and cognitive medical bias against the elderly, resulting in inadequate surgery, suboptimal chemotherapy, and poor enrollment in clinical trials [4].

Improving the prognosis of elderly patients with gynecological cancers requires a better understanding of the differences in tumor biology between younger and elderly patients and methods of distinguishing between patients who can and cannot tolerate standard cytoreductive surgery and chemotherapy. Not all of the elderly cancer patients we manage can participate in conventional clinical trials and many are vulnerable elderly patients. Differences also exist in the value of outcomes sought from cancer treatment between elderly cancer patients and younger patients. Not only do objective outcomes such as prolonged survival and adverse events need to be evaluated, but also subjective outcomes such as changes in quality of life (QOL). Currently, patient-reported outcomes, in which patients themselves report on treatment effects and QOL without interpretation by clinicians or others, are gaining popularity [5, 6]. To reduce the disadvantages of treatment for vulnerable patients, changes in the appropriateness of surgical treatment, chemotherapy dose, schedule, and timing (neoadjuvant or postoperative) may need to be considered.

## 9.2 Current Status and Problems of Surgical Treatment for Elderly Gynecological Cancer Patients

An analysis of more than 12,000 patients found that ovarian cancer patients over 80 years old were less likely to undergo surgery and less likely to undergo optimal tumor reduction procedures [7]. In addition, Fairfield et al. reported regional differences in ovarian cancer mortality among Medicare patients, and those with poor access to facilities that can perform surgery for cancer treatment may not be adequately treated, which may contribute to the poor outcomes seen in elderly women [8].

However, aggressive primary surgical cytoreduction in elderly patients is clearly quite invasive, according to Surveillance, Epidemiology, and End Results (SEER) Medicare analysis. Among women ≥65 years old with stage III or IV ovarian cancer who underwent primary cytoreduction, the 30-day mortality rate was 5.6% among those admitted on a waitlist and 20.1% among patients admitted on an emergency basis. In contrast, patients >75 years old with stage III or IV disease and one or more comorbidities showed a 30-day mortality rate of 12.7% even with elective hospitalization [9]. A further concern is that the degree of surgical invasiveness may result in patients not receiving additional adjuvant chemotherapy. In one retrospective report of 85 patients ≥80 years old who underwent tumor reduction surgery (primarily primary surgery), 13% died before discharge and 20.1% died within 60 days of surgery. Thirteen percent received no adjuvant therapy, and of those who did receive chemotherapy, 43% received three or fewer cycles of treatment [10]. These and other results illustrate that the more invasive nature of primary cytoreductive surgery in elderly patients has led to increased use of neoadjuvant chemotherapy and interval cytoreductive surgery in the elderly [11].

Although older cancer patients often show poor prognosis, some patients can achieve a status of no residual disease with cytoreductive surgery. The Asian Oncological Group found that the percentage of patients in clinical trials with no residual disease after primary surgery was 45.1% in patients <50 years old, 25.7% in patients 50–64 years old, and 25.7% in patients ≥65 years old [12]. In a series of 280 consecutive patients ≥65 years old who underwent primary surgery at the Mayo Clinic, the percentage with residual lesions >1 cm was 43% among women ≥80 years old and 25% in women 65–69 years old. The 3-month mortality rate was 25% in women ≥80 years old and 4% in women 65–69 years old [13].

Thus, even in the same elderly population, accurate assessment is needed before surgery to clarify which patients deserve aggressive tumor reduction surgery followed by standard chemotherapy and which patients should be offered alternative therapies such as neoadjuvant chemotherapy followed by interval cytoreductive surgery or primary chemotherapy alone.

## 9.3 Current Status of Problems with Chemotherapy for Elderly Gynecological Cancer Patients

Elderly women are less likely to receive any chemotherapy, let alone standard chemotherapy. A SEER Medicare analysis found that among women ≥65 years old diagnosed with ovarian cancer between 2001 and 2005, 29% did not receive any chemotherapy, 25% received only partial chemotherapy, and only 47% completed the scheduled chemotherapy. In addition, patients ≥80 years old were twice as likely not to complete chemotherapy, and patients with two or more comorbidities were 83% more likely not to complete chemotherapy. Those results were considered to indicate that chemotherapy may be underused in older women, and that high-level retrospective analysis cannot determine whether "underuse" of chemotherapy was actually medically appropriate [14].

Observational studies have shown that first-line chemotherapy improves survival in elderly women with ovarian cancer. Unfortunately, only half of this population receive platinum-based chemotherapy, according to SEER demographics, which included approximately 8000 women ≥65 years old with stage III or IV epithelial ovarian cancer [15]. A multivariate review that took into account cancer type, comorbidities, and other factors suggested that patients who received surgery alone displayed similar survival rates compared with those who received no treatment (22 months vs. 17 months, respectively), but patients who received chemotherapy as the only treatment achieved prolonged overall survival (14.4 months) [16].

The most common toxicities of platinum-taxanes, as the usual first-line agents for ovarian cancer, are cytopenia and neuropathy. This was demonstrated in a large retrospective analysis of the outcomes and toxicities in 620 patients ≥70 years old enrolled in GOG 182, a phase III trial investigating triple therapy in patients with newly diagnosed ovarian cancer [4]. Although elderly women enrolled in such trials were more likely to be healthy than the average older ovarian cancer patient, older patients still showed lower performance status (PS), lower rates of completion of all eight chemotherapy cycles, and increased toxicity, particularly grade 3 neutropenia and grade 2 or higher neurological disability (36 vs. 20% for younger women in the standard carboplatin/paclitaxel arm).

Although the difference in median time to disease progression was only 1 month, elderly women displayed significantly shorter median overall survival (37 months) than younger women (45 months, $P < 0.001$). This shows that elderly patients are more vulnerable to the toxicities of certain chemotherapies.

As previously mentioned, age is a strong predictor of survival in ovarian cancer and often influences treatment strategy. The basis of cancer treatment is surgery and chemotherapy. To improve the utility and tolerability of treatment (surgery and chemotherapy) for elderly patients with gynecological cancer, better geriatric assessment tools specific to the elderly need to be developed.

## 9.4 Personalized Assessment of Treatment for Elderly Cancer Patients

In cancer treatment, elderly patients are considered "fit" (a condition in which the patient can receive the same standard treatment as a younger adult) or "unfit" (a condition in which the patient is unable to receive the same standard treatment as a younger adult). To evaluate the suitability or unsuitability of highly invasive treatment for "unfit" elderly cancer patients, the following classifications should be used: "Frail patients: Patients whose condition is considered unsuitable for aggressive cancer treatment. Vulnerable patients: Patients considered ineligible for active cancer treatment, i.e., the condition is so bad that the patient cannot expect or tolerate treatment." Not only is determining which stage of disease the patient is important (i.e., "the patient cannot receive the same standard treatment as healthy adults, but can receive less intense treatment or treatment with less toxicity"), but also discussing this with the patient and family in order to decide on a treatment plan [17] (Fig. 9.1).

Although there is no standard tool for differentiating between fit and unfit patients that can be used to examine common indicators for all cancer types, the elderly vary greatly from person to person, and determination of a medical

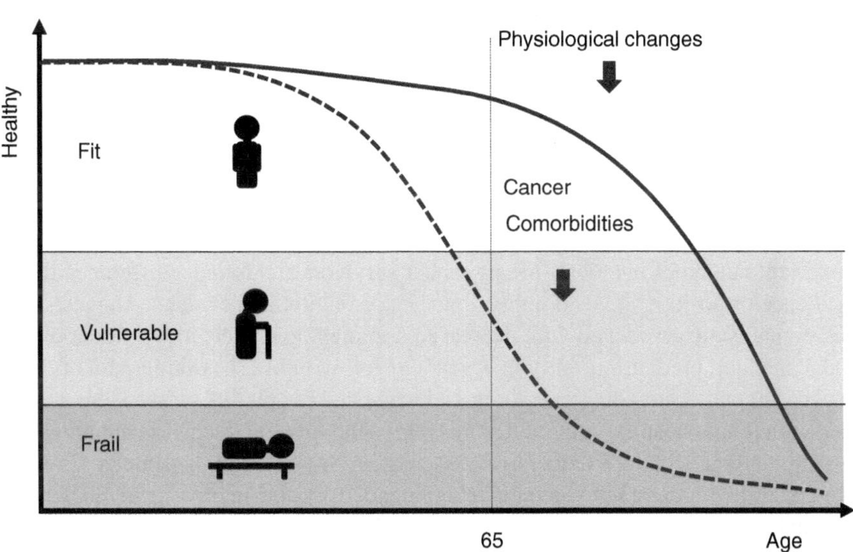


**Fig. 9.1** Conceptual division of the target population when treatment is performed. Fit: A condition in which invasive treatment is available as standard treatment. Frail: A condition not meeting the indications for active cancer treatment. Vulnerable: A condition in which the standard treatment for healthy adults cannot be provided, but less toxic treatment can be applied

treatment policy must be based on a comprehensive evaluation of physical and mental disabilities, social and economic problems, and calendar age. Particularly in cases of cognitive impairment, confirmation of the patient's will becomes an issue in determining the medical treatment policy. Comprehensive geriatric assessment (CGA) is a patient assessment method that focuses on these issues.

CGA is a multidimensional assessment tool that provides a complete picture of the elderly individual. CGA is a tool for comprehensively assessing daily living functions, including activities of daily living (ADL), instrumental ADL (IADL), cognitive function, mood, social factors, and home environment, in addition to disease assessment. CGA in older cancer patients includes assessment of functional status, comorbid medical conditions, cognition, nutritional status, psychological status, social support, and medication status. In the field of geriatric oncology, CGA is a concept that combines assessment and intervention. In oncology, such assessment is described as geriatric assessment (GA), with the implication that only a comprehensive assessment is performed to determine treatment strategy [17, 18].

## 9.5 Personalized Treatment Based on Preoperative Assessment in the Elderly

GA is undoubtedly a well-established method for assessing the elderly and optimizing preoperative diagnostic and treatment planning. A Cochrane meta-analysis of 22 clinical trials of more than 10,000 emergency admissions comparing GA with standard care found that patients who underwent GA assessment displayed improved survival and home discharge rates at both 6- and 12-month follow-ups [19].

Robinson et al. reported a difference in the incidence of postoperative complications among 72 patients with colorectal cancer classified as fit, vulnerable, or frail [20]. Clough-Gorr et al. evaluated several GA domains associated with clinically important outcomes in elderly breast cancer survivors and found a difference in the incidence of postoperative complications [21]. Longitudinal evidence suggests that GA domains are associated with decreased treatment tolerance, independent of age and stage, and predict mortality at 7 years of follow-up [22]. Among patients who underwent pancreaticoduodenectomy, older age and poorer GA scores were associated with longer hospital stays and admission to the surgical intensive care unit [23]. In surgical practice, the Preoperative Assessment of Cancer in the Elderly (PACE), a preoperative assessment method recommended by the International Society of Geriatric Oncology (SIOG), which is based on the GA, the Physiological & Operative Severity Score for PACE, which integrates GA, the Physiological & Operative Severity Score for numeration of Mortality and Morbidity (POSSUM), the Portsmouth modification (P-POSSUM), and the American Society of Anesthesiologists (ASA) physical status classification system are valuable tools for identifying vulnerable elderly cancer patients. These tools can also reduce inappropriate inequities due to age in patients scheduled to undergo surgery [24].

However, GA requires experience, is time consuming to assess, and is not necessary for all patients. Several screening methods have thus been developed.

**Table 9.1** List of screening methods for CGA commonly used in oncology for elderly patients

| Test | Patients | Number of domains | Range | Cut-off scores | Ref. |
|---|---|---|---|---|---|
| VES-13 | General | 13 | 0–15 | ≥3 | Saliba S |
| TRST | Emergency | 5 | 0–6 | ≥1 | Meldon SW |
| G8 | Oncology | 8 | 0–17 | ≤14 | Soubeyran P |
| GFI | General | 15 | 0–15 | ≥4 | Slaets JP |
| aCGA | Oncology | 15 | ADL: 3<br>IAD: 4<br>GDS: 4<br>MMS: 4 | ≥1 dependence<br>≥1 dependence<br>≥ 2<br>≤ 6 | Overcash JA |
| Rockwood | General | 4 | 0–3 | ≥2 | Rockwood K |
| Balducci | General | 4 | 0–4 | 1 | Balducci L |
| Fried score | General | 5 | 0–5 | ≥3 | Fried LP |

Abbreviations: *VES-13* Vulnerable Elderly Survey-13, *TRST* Triage Risk Screening Tool, *GFI* Groningen Frailty Index, *aCGA* abbreviated Comprehensive Geriatric Assessment, *ADL* Activities of Daily Living score, *IADL* Instrumental Activities of Daily Living score, *GDS* Geriatric Depression Scale, *MMS* Mini Mental Health Status

**Table 9.2** Overview of the diagnostic efficiency of screening tests

| Test | Sensitivity range (%) | Specificity range (%) | PPV range (%) | NPV range (%) |
|---|---|---|---|---|
| VES-13 | 39–88 | 62–100 | 65–100 | 48–52 |
| TRST | 64–91 | 42–100 | 81–100 | 47–63 |
| G8 | 77–92 | 52–75 | 78–86 | 61–78 |
| GFI | 39–66 | 69–87 | 86–90 | 40–59 |
| aCGA | 51 | 97 | 97 | 43 |
| Rockwood | 47 | 88 | – | – |
| Balducci | 94 | 50 | – | – |
| Fried score | 23–37 | 86–96 | 77 | 66 |

A dash indicates the data were not available
Diagnostic efficiency of screening methods for CGA commonly used in oncology
*PPV* positive predictive value, *NPV* negative predictive value. Studies that did not provide confidence intervals were excluded from the studies reviewed

Alternatives to the full GA that are commonly used in oncology include Fried's Flail Diagnostic Criteria [25], Balducci's Diagnostic Criteria [26], the Vulnerable Elderly Survey (VES-13) [27], Triage Risk Screening Tool [28], Geriatric 8 (G8) [29], Groningen Frailty Index [30], abbreviated Comprehensive Geriatric Assessment [31], and Rockwood [32]. Screening methods and the reported diagnostic efficiency for each method are shown in Table 9.1.

According to reported meta-analyses, the median prevalence of vulnerable status based on screening scores is 49% (range, 12–83%). On the other hand, the prevalence of pre-frail status based on GA is 68% (range, 28–94%) [33–37] (Table 9.2). However, some caution must be exercised in interpreting the results of the above studies. First, many articles have included cancer patients undergoing chemotherapy, and few have examined patients scheduled for surgery. Furthermore, the domains of the reference Gas used varied, and the variability in Gas presented must be taken into account. Most studies have considered at least three of the following domains: cognitive functioning, mood and depression, nutritional status, ADL, IADL, comorbidity, polypharmacy, mobility, and social support. However, in order to consider problematic areas, each domain of GA and the cutoff used for this purpose may differ. These methodological differences significantly complicate comparisons with GA results; the SIOG states that Gas, at least for elderly cancer patients, should include assessment domains for functional status, cognition, and mood [38]. In addition, it should be noted that many of the published articles do not directly evaluate patients considered for surgery, present GA before adjuvant treatment, or mix patients with different malignancies (gastrointestinal, breast, and hematological cancers), and some studies do not mention the type of treatment [33–37].

For the general cancer patient population, the screening method offering the highest sensitivity is the G8 [33–37]. The G8 is based on the Mini Nutritional Assessment (MNA). Among the GA domains, the accuracy of the MNA in predicting frailty is about 80%, and a significant correlation between the two variables has been observed [29]. However, specificity is not as high for G8, at 39–75% [33–37]. One of the most important characteristics of the screening tool, however, is exclusion of the possibility of vulnerable status, corresponding to a negative predictive value (NPV). G8 shows a high NPV, but includes subjects other than those scheduled for cancer surgery. Kenis et al. found that in 937 patients with various cancers, G8 offered high sensitivity (84%) and NPV (91%) but low specificity (31% for both ADL and IADL) for functional declines in ADL and IADL. Elderly patients with normal G8 are at lower risk of functional declines in ADL and less pronounced declines in IADL. Patients with non-normal G8 require further evaluation during extended follow-up [35]. The same group of investigators also examined the occurrence of adverse events during treatment in the above groups of patients. G8 showed significant prognostic improvement for OS (median survival: 31.8 months in abnormal G8; not reached in normal G8 [hazard ratio 0.38; 95% confidence interval 0.27–0.52; $P < 0.001$]). Elderly patients with normal G8 displayed a 62% lower chance of death after a median follow-up of 18.95 months [36]. Liuu et al. examined the survival of 518 patients with various cancers. In multivariate analysis, G8 abnormalities were associated with increased 6-month mortality (hazard ratio 6.68; 95% confidence interval 1.63–27.35; $P = 0.001$) [39].

NRG Oncology devised their own GA-GYN score. This was an attempt to predict postoperative prognosis by scoring: 1 point for IADLs requiring medication management, 2 points for ADLs with limited mobility, 1 point for occasional deterioration in high-level ADL due to physical or mental reasons, 1 point if the patient had experienced a fall in the past 6 months, 2 points if the patient was deaf, 2 points if the patient was >72 years old, 3 points for hemoglobin level <10 g/dl, and 3 points for creatinine clearance <34 ml/min. Preoperative GA-GYN assessment did not prove useful in predicting the occurrence of serious complications after ovarian cancer surgery. However, the score was associated with the occurrence of serious postoperative complications in patients who underwent stage III–IV open cytoreduction surgery [40]. Further studies of preoperative assessment for gynecological diseases are needed.

## 9.6 Personalized Treatment Based on Pre-chemotherapy Assessment in the Elderly

For chemotherapy in the elderly, the American Society of Clinical Oncology published a guideline on "Practical assessment and interventions for vulnerable cancer patients starting chemotherapy" in 2018. The guideline comprises four major points. (1) When chemotherapy is given to patients ≥65 years old, GA (assessment of physical function, physical performance and risk of falls, comorbidities, depression, social activities/support, nutritional status, and cognitive function) should be used to identify vulnerabilities. (2) The Expert Panel recommended the following GA tools for their usefulness and ease of use in predicting adverse events: (a) at a minimum, assess physical function, comorbidity, falls, depression, cognitive function, and nutrition; (b) IADL (physical functioning), careful history taking or assessment tools (comorbidities), questions about falls, Geriatric Depression Scale (depression), Mini-Cog or Blessed Orientation-Memory-Concentration (BOMC) test (cognitive functioning), and weight loss (nutrition); or (c) Chemotherapy Risk Assessment Scale for High-Age Patients (CRASH) score [41] (Tables 9.3 and 9.4) or Cancer Aging Research Group (CARG) toxicity scores [42] (Table 9.5). The

**Table 9.3** Chemotherapy Risk Assessment Scale for High-Age Patients (CRASH) score

| | Scores | | |
|---|---|---|---|
| Predictors | 0 | 1 | 2 |
| *Hematologic score* (predictors of Grade 4 hematologic toxicity) | | | |
| Diastolic BP (mmHg) | ≤72 | >72 | |
| IADL | 26–29 | 10–25 | |
| LDH (if ULN 618 U/L; otherwise, 0.74/L*ULN) | 0–459 | | >459 |
| Chemotox | 0 | 1 | 2 |
| *Nonhematologic score* (predictors of grade 3/4 nonhematologic toxicity) | | | |
| ECOG PS | 0 | 1–2 | 3–4 |
| MMS | 30 | | <30 |
| MNA | 28–30 | | <28 |
| Chemotox | 0 | 1 | 2 |

Abbreviations: *BP* blood pressure, *Chemotox* toxicity of the chemotherapy regimen (see Table 9.4), *ECOG PS* Eastern Cooperative Oncology Group performance status, *IADL* Instrumental Activities of Daily Living score, *LDH* lactate dehydrogenase, *MMS* Mini Mental Health Status, *MNA* Mini Nutritional Assessment, *ULN* upper limit of normal

CRASH score, an evaluation tool for predicting adverse reactions to chemotherapy in elderly cancer patients. (1) Predictors of grade 4 hematological toxicity are associated with instrumental activities of daily living (IADL), lactate dehydrogenase (LDH), diastolic blood pressure (BP), and the published toxicity of anticancer drugs (Chemotox) (risk categories: low (0–1 points), 7%; medium-low (2–3 points), 23%; medium-high (4–5 points), 54%; and high (≥6 points), 100%, respectively; $P < 0.001$). (2) Predictors of grade 3/4 non-hematological toxicity are associated with malnutrition (Mini Nutritional Assessment score; MNA), cognition (Mini-Mental State Examination score; MMSE), Eastern Cooperative Oncology Group performance status (ECOG PS) score, and Chemotox (risk categories: low (0–2 points), 33%; medium-low (3–4 points), 46%; medium-low (5–6 points), 67%; and high (≥6 points), 93%, respectively; $P < 0.001$). Combined risk categories are 50%, 58%, 77%, and 79%, respectively ($P < 0.001$)

**Table 9.4** Anticancer drug toxicity (Chemotox) score

| CRASH score | | |
|---|---|---|
| 0 | 1 | 2 |
| Docetaxel weekly | Carboplatin/gemcitabine AUC 4-6/1 g days 1 and 8 | Carboplatin/docetaxel q3w |
| Paclitaxel weekly | Carboplatin/paclitaxel q3w | Cisplatin/docetaxel 75/75 |
| Gemcitabine 1g q3w | Cisplatin/gemcitabine days 1 and 8 | Cisplatin/gemcitabine days 1, 8 and 15 |
| Gemcitabine 1.25g q3w | Gemcitabine q7w then q3w | Cisplatin/paclitaxel 135-24 h q3w |
| Dacarbazine | Gemcitabine/irinotecan | Paclitaxel q3w |
| | PEG doxorubicin 50 mg q4w | Docetaxel q3w |
| | Topotecan weekly | Doxorubicin q3w |
| | | Irinotecan q3w |
| | | Topotecan monthly |

Abbreviations: *CRASH* Chemotherapy Risk Assessment, *AUC* area under the concentration-time curve, *PEG* pegylated, *q3w* every 3 weeks, *q4w* every 4 weeks
Toxicity of the chemotherapy regimen as calculated using the MAX2 method. The MAX2 index is the average of the most frequent grade 4 hematological toxicities and the most frequent grade 3–4 non-hematological toxicities as reported in publications of a regimen. This index correlates well with the average overall risk of severe toxicity for that regimen. Boundaries between points are: MAX2 < 0–0.44 = 0; 0.45–0.57 = 1; >0.57 = 2. A list of scores is available online (https://moffitt.org/eforms/crashscoreform/; accessed March 5, 2021). This figure shows only the most commonly used regimens, particularly for gynecological cancer patients

**Table 9.5** CARG toxicity score

| Risk factor | | Score 0 | 1 | 2 | 3 |
|---|---|---|---|---|---|
| Age (years) | | <72 | | ≥72 | |
| Cancer type | | Others | | GI or GU | |
| Chemotherapy dosing, standard dose | | Reduced | | Standard | |
| No. of chemotherapy drugs, polychemotherapy | | No | | Yes | |
| Hemoglobin (g/dl) | Male | ≥11 | | | <11 |
| | Female | ≥10 | | | <10 |
| Creatinine clearance (Jelliffe, ideal weight) (ml/min) | | ≥34 | | | <34 |
| Hearing, fair or worse | | No | | Yes | |
| No. of falls in last 6 months, 1 or more | | None | | | ≥1 |
| IADL: taking medications, with some help/unable | | No | Yes | | |
| MOS: walking 1 block, somewhat limited/limited a lot | | No | | Yes | |
| MOS: decreased social activity because of physical/emotional health, limited at least sometimes | | No | Yes | | |

Abbreviations: *GU* genitourinary, *IADL* instrumental activities of daily living, *MOS* Medical Outcomes Study
CARG score (an evaluation tool for predicting adverse reactions to chemotherapy in elderly cancer patients) data from 500 elderly individuals added to the CRASH database. Risk scores are divided into three categories based on the risk of grade 3–5 toxicity: low risk, 0–5 points; intermediate risk, 6–9 points; and high risk, 10–19 points. The low-risk group comprises 70% of total patients with a 30% incidence of grade 3–5 adverse events, the intermediate-risk group comprises 48% of patients with a 52% incidence of adverse events, and the high-risk group comprises 17% of patients with an 83% of incidence of adverse events. Toxicity profiles differ significantly among risk groups ($P < 0.001$)

CARG is a total risk score derived from a GA consisting of 11 predictors of serious adverse events. This score can predict the toxicity of chemotherapy, which is rarely predicted by conventional measures of PS. CRASH score is characterized by separate analyses of hematological and non-hematological toxicities, including Hematological score, non-Hematological score, and Total score. A high correlation is seen between Hematological score and the incidence of hematological toxicities in G4. As for other Gas, G8 or VES-13 is used to predict prognosis. (3) Clinicians should use one of the validated tools described in ePrognosis to estimate life expectancy over 4 years. (4) The process of managing elderly cancer patients by GA as a consensus of experts using the Delphi method is recommended: (a) use the GA results to predict the risk of adverse events to develop an individualized treatment plan, identify non-cancer problems, and intervene; (b) share the GA results with patients and their families to develop a treatment plan; and (c) share GA results with patients and their families to assist treatment decision-making, and provide GA-based interventions for non-oncology problems [43].

Several reports have described clinical trials using GA in cancer pharmacotherapy for gynecological cancer. A clinical research group at GINECO in France developed a new prognostic tool in the treatment of cancer in the elderly. The Geriatrics Vulnerability Score (GVS) consists of ADL, IADL, albumin level, lymphocyte count, and HADS score. A positive score in three of these items is associated with a lower treatment completion rate, increased grade 3/4 non-hematological toxicity, and unexpected hospitalization [44]. Recently, the same group conducted a prospective, randomized clinical trial in 120 elderly, vulnerable patients with ovarian cancer. The study compared the feasibility, efficacy, and safety of single-agent carboplatin every 3 weeks, weekly carboplatin and paclitaxel, and conventional carboplatin and paclitaxel every 3 weeks. The results showed that single-agent carboplatin was associated with lower feasibility and activity than carboplatin and paclitaxel, completing six cycles is 48%, 60%, and 65%, respectively, and significantly worse progression-free and overall survivals compared to the conventional 3-week carboplatin-paclitaxel regimen, leading to early termination of the trial [45]. An Italian group conducted a prospective clinical trial to determine whether the VES-13, as a comprehensive assessment of the elderly, could predict the prognosis of elderly gynecological cancer patients >70 years old. They reported that 42.9% of patients were considered vulnerable elderly, and that serious hematological and non-hematological toxicities were frequently observed in this group, leading to discontinuation of drug reduction [46].

## 9.7 Summary of the Basic Concept of Individualized Medicine in the Treatment of Elderly Cancer Patients

1. Do not provide highly invasive treatment to patients unwilling to receive treatment, or to patients capable of making treatment decisions who refuse treatment after presenting various treatment options. In addition, the content and intensity of supportive and palliative care should be discussed with the patient and family regarding the level to which these should be implemented.

2. Regardless of age, appropriate medical care should be provided to fit patients who are eligible and able to receive standard treatment. Efforts should be made to avoid both under- and overtreatment. For this purpose, multidisciplinary care involving local medical institutions is required.
3. For vulnerable patients who are eligible for treatment but for whom standard treatment is difficult to implement, less-invasive surgical techniques with fewer side effects and reduced treatment intensity (dose intensity, irradiation field, and radiation dose) should be selected over standard treatment, to achieve a balance between efficacy and adverse events, and to allow flexible response.
4. Conservative medical care should be provided for patients who have difficulty with aggressive treatment (frail patients).
5. Supportive and palliative care should be based on the same concept, with active treatment for fit and vulnerable patients, and symptom relief for frail patients. In other words, adverse events caused by supportive and palliative medicine should be avoided, as should overtreatment of patients who are at the end-of-life stage.
6. Our ultimate goal is for all cancer patients, young and old, to achieve a cure with good QOL in the curable stage, and that even in the difficult-to-cure stage, patients can live out the rest of their lives with minimal pain by making full use of supportive and palliative medicine while receiving various medical interventions (irrespective of fit, vulnerable or frail status).

**Acknowledgments** None

**Disclosure** None of the authors have any conflicts of interest to declare in relation to this chapter.

## References

1. Oberaigner W, Minicozzi P, Bielska-Lasota M, Allemani C, de Angelis R, Mangone L, Sant M, Eurocare Working Group. Survival for ovarian cancer in Europe: the across-country variation did not shrink in the past decade. Acta Oncol. 2012;51(4):441–53.
2. Yancik R, Ries LA. Cancer in older persons: an international issue in an aging world. Semin Oncol. 2004;31(2):128–36.
3. Sant M, Chirlaque Lopez MD, Agresti R, Sánchez Pérez MJ, Holleczek B, Bielska-Lasota M, Dimitrova N, Innos K, Katalinic A, Langseth H, Larrañaga N, Rossi S, Siesling S, Minicozzi P, EUROCARE-5 Working Group. Survival of women with cancers of breast and genital organs in Europe 1999–2007: results of the EUROCARE-5 study. Eur J Cancer. 2015;51(15):2191–205.
4. Tew WP, Muss HB, Kimmick GG, Von Gruenigen VE, Lichtman SM. Breast and ovarian cancer in the older woman. J Clin Oncol. 2014;32(24):2553–61. https://doi.org/10.1200/JCO.2014.55.3073.
5. Kluetz P, Chingos DT, Basch EM, et al. Patient-reported outcomes in cancer clinical trials: measuring symptomatic adverse events with the National Cancer Institute's patient-reported outcomes version of the common terminology criteria for adverse events (PRO-CTCAE). Am Soc Clin Oncol Educ Book. 2016;35:67–73.
6. LeBlanc TW, Abernethy AP. Patient-reported outcomes in cancer care—hearing the patient voice at greater volume. Nat Rev Clin Oncol. 2017;14(12):763–72.

7. Hightower RD, Nguyen HN, Averette HE, Hoskins W, Harrison T, Steren A. National survey of ovarian carcinoma. IV: Patterns of care and related survival for older patients. Cancer. 1994;73(2):377–83.
8. Fairfield KM, Lucas FL, Earle CC, Small L, Trimble EL, Warren JL. Regional variation in cancer-directed surgery and mortality among women with epithelial ovarian cancer in the Medicare population. Cancer. 2010;116(20):4840–8.
9. Thrall MM, Goff BA, Symons RG, Flum DR, Gray HJ. Thirty-day mortality after primary cytoreductive surgery for advanced ovarian cancer in the elderly. Obstet Gynecol. 2011;118(3):537–47.
10. Moore KN, Reid MS, Fong DN, Myers TK, Landrum LM, Moxley KM, Walker JL, McMeekin DS, Mannel RS. Ovarian cancer in the octogenarian: does the paradigm of aggressive cytoreductive surgery and chemotherapy still apply? Gynecol Oncol. 2008;110(2):133–9.
11. Thrall MM, Gray HJ, Symons RG, Weiss NS, Flum DR, Goff BA. Neoadjuvant chemotherapy in the Medicare cohort with advanced ovarian cancer. Gynecol Oncol. 2011;123(3):461–6. https://doi.org/10.1016/j.ygyno.2011.08.030.
12. Wimberger P, Lehmann N, Kimmig R, Burges A, Meier W, Hoppenau B, du Bois A, AGO-OVAR. Impact of age on outcome in patients with advanced ovarian cancer treated within a prospectively randomized phase III study of the Arbeitsgemeinschaft Gynaekologische Onkologie Ovarian Cancer Study Group (AGO-OVAR). Gynecol Oncol. 2006;100(2):300–7.
13. Langstraat C, Aletti GD, Cliby WA. Morbidity, mortality and overall survival in elderly women undergoing primary surgical debulking for ovarian cancer: a delicate balance requiring individualization. Gynecol Oncol. 2011;123(2):187–91.
14. Hershman D, Jacobson JS, McBride R, Mitra N, Sundararajan V, Grann VR, Neugut AI. Effectiveness of platinum-based chemotherapy among elderly patients with advanced ovarian cancer. Gynecol Oncol. 2004;94(2):540–9. https://doi.org/10.1016/j.ygyno.2004.04.022.
15. Lin JJ, Egorova N, Franco R, Prasad-Hayes M, Bickell NA. Ovarian cancer treatment and survival trends among women older than 65 years of Age in the United States, 1995–2008. Obstet Gynecol. 2016;127(1):81–9. https://doi.org/10.1097/AOG.0000000000001196.
16. Fairfield KM, Murray K, Lucas FL, Wierman HR, Earle CC, Trimble EL, Small L, Warren JL. Completion of adjuvant chemotherapy and use of health services for older women with epithelial ovarian cancer. J Clin Oncol. 2011;29(29):3921–6.
17. ESMO. Handbook of cancer in the senior patient. New York: Informa; 2010.
18. Wildider H. International Society of Geriatric Oncology consensus on geriatric assessment in older patients with cancer. J Clin Oncol. 2014;32:2595–603.
19. Ellis G, Whitehead MA, O'Neill D, et al. Comprehensive geriatric assessment for older adults admitted to hospital. Cochrane Database Syst Rev. 2011;7:CD006211.
20. Robinson TN, Wu DS, Pointer L, et al. Simple frailty score predicts postoperative complications across surgical specialties. Am J Surg. 2013;206:544–50.
21. Clough-Gorr KM, Stuck AE, Thwin SS, et al. Older breast cancer survivors: geriatric assessment domains are associated with poor tolerance of treatment adverse effects and predict mortality over 7 years of follow-up. J Clin Oncol. 2010;28:380–6.
22. Giannotti C, Sambuceti S, Signori A, et al. Frailty assessment in elective gastrointestinal oncogeriatric surgery: predictors of one-year mortality and functional status. J Geriatr Oncol. 2019;10:716–23.
23. Dale W, Hemmerich J, Kamm A, et al. Geriatric assessment improves prediction of surgical outcomes in older adults undergoing pancreaticoduodenectomy: a prospective cohort study. Ann Surg. 2014;259:960–5.
24. PACE Participants, Audisio RA, Pope D, Ramesh HS, et al. Shall we operate? Preoperative assessment in elderly cancer patients (PACE) can help. A SIOG surgical task force prospective study. Crit Rev Oncol Hematol. 2008;65:156–63.
25. Fried LP, Tangen CM, Walston J, et al. Frailty in older adults: evidence for a phenotype. J Gerontol A Biol Sci Med Sci. 2001;56(M146–56):6.

26. Balducci L, Stanta G. Cancer in the frail patient. A coming epidemic. Hematol Oncol Clin North Am. 2000;14:235–50.
27. Saliba S, Elliott M, Rubenstein LA, et al. The Vulnerable Elders Survey (VES-13): a tool for identifying vulnerable elders in the community. JAGS. 2001;49:1691–9.
28. Meldon SW, Mion LC, Palmer RM, et al. A brief risk-stratification tool to predict repeatemergency department visits and hospitalizations in older patients discharged from the emergency department. Acad Emerg Med. 2003;10:224–32.
29. Soubeyran P, Bellera C, Goyard J, et al. Validation of the G8 screening tool in geriatric oncology: the ONCODAGE project. J Clin Oncol. 2011;29(suppl; abstr 9001;2011 ASCO Annual Meeting).
30. Slaets JP. Vulnerability in the elderly: frailty. Med Clin North Am. 2006;90:593–601.
31. Overcash JA, Beckstead J, Moody L, et al. The abbreviated comprehensive geriatric assessment (aCGA) foruse in the older cancer patient as a prescreen: scoring andinterpretation. Crit Rev Oncol Hematol. 2006;59:205–10.
32. Rockwood K, Stadnyk K, MacKnight C, McDowell I, Hebert DBR. A brief clinical instrument to classify frailty in elderlypeople. Lancet. 1999;353:205–6.
33. Decoster L, Van Puyvelde K, Mohile S, et al. Screening tools for multidimensional health problems warranting a geriatric assessment in older cancer patients: an update on SIOG recommendations. Ann Oncol. 2015;26:288–300.
34. Puts MT, Hardt J, Monette J, Girre V, Springall E, Alibhai SM, et al. Use of geriatric assessment for older adults in theoncology setting: a systematic review. JNCI J Natl Cancer Inst. 2012;104:1134–64.
35. Kenis C, Bron D, Libert Y, et al. Relevance of a systematic geriatric screening and assessment in older patients with cancer: results of a prospective multicentric study. Ann Oncol. 2013;24:1306–12.
36. Hamaker ME, Jonker JM, de Rooij SE, et al. Frailty screening methods for predicting outcome of a comprehensive geriatric assessment in elderly patients with cancer: a systematic review. Lancet Oncol. 2012;13:37–44.
37. Owusu C, Koroukian SM, Schluchter M, et al. Screening older cancer patients for a Comprehensive Geriatric Assessment: a comparison of three instruments. J Geriatr Oncol. 2011;2:121–9.
38. Extermann M, Aapro M, Bernabei R, et al. Use of comprehensive geriatric assessmentin older cancer patients: recommendations from the task forceon CGA of the International Society of Geriatric Oncology (SIOG). Crit Rev Oncol Hematol. 2005;55:241–52.
39. Liuu E, Canouï-Poitrine F, Tournigand C, Laurent M, Caillet P, Le Thuaut A, Vincent H, Culine S, Audureau E, Bastuji-Garin S, Paillaud E, ELCAPA Study Group. Accuracy of the G-8 geriatric-oncology screening tool for identifying vulnerable elderly patients with cancer according to tumour site: the ELCAPA-02 study. J Geriatr Oncol. 2014;5(1):11–9. https://doi.org/10.1016/j.jgo.2013.08.003.
40. Ahmed A, Deng W, Tew W, Bender D, Mannel RS, Littell RD, DeNittis AS, Edelson M, Morgan M, Carlson J, Darus CJ, Fleury AC, Modesitt S, Olawaiye A, Evans A, Fleming GF. Pre-operative assessment and post-operative outcomes of elderly women with gynecologic cancers, primary analysis of NRG CC-002: an NRG oncology group/gynecologic oncology group study. Gynecol Oncol. 2018;150(2):300–5.
41. Extermann M, et al. Predicting the risk of chemotherapy toxicity in older patients: the Chemotherapy Risk Assessment Scale for High-Age Patients (CRASH) score. Cancer. 2012;118:3377–86.
42. Hurria A, Togawa K, Mohile SG, et al. Predicting chemotherapy toxicity in older adults with cancer: a prospective multicenter study. J Clin Oncol. 2011;29:3457–65.
43. Mohile SG, Dale W, Somerfield MR, Hurria A. Practical assessment and management of vulnerabilities in older patients receiving chemotherapy: ASCO guideline for geriatric oncology summary. J Oncol Pract. 2018;14(7):442–6. https://doi.org/10.1200/JOP.18.00180.

44. Falandry C, Weber B, Savoye AM, Tinquaut F, Tredan O, Sevin E, Stefani L, Savinelli F, Atlassi M, Salvat J, Pujade-Lauraine E, Freyer G. Development of a geriatric vulnerability score in elderly patients with advanced ovarian cancer treated with first-line carboplatin: a GINECO prospective trial. Ann Oncol. 2013;24(11):2808–13.
45. Falandry C, Rousseau F, Mouret-Reynier MA, Tinquaut F, Lorusso D, Herrstedt J, Savoye AM, Stefani L, Bourbouloux E, Sverdlin R, D'Hondt V, Lortholary A, Brachet PE, Zannetti A, Malaurie E, Venat-Bouvet L, Trédan O, Mourey L, Pujade-Lauraine E, Freyer G, Groupe d'Investigateurs Nationaux pour l'Étude des Cancers de l'Ovaire et du sein (GINECO). Efficacy and safety of first-line single-agent carboplatin vs carboplatin plus paclitaxel for vulnerable older adult women with ovarian cancer: a GINECO/GCIG randomized clinical trial. JAMA Oncol. 2021;7(6):853–61. https://doi.org/10.1001/jamaoncol.2021.0696.
46. Ferrero A, Villa M, Tripodi E, Fuso L, Menato G. Can Vulnerable Elders Survey-13 predict the impact of frailty on chemotherapy in elderly patients with gynaecological malignancies? Medicine (Baltimore). 2018;97(39):e12298. https://doi.org/10.1097/MD.0000000000012298.

GPSR Compliance

*The European Union's (EU) General Product Safety Regulation (GPSR) is a set of rules that requires consumer products to be safe and our obligations to ensure this.*

*If you have any concerns about our products, you can contact us on ProductSafety@springernature.com*

In case Publisher is established outside the EU, the EU authorized representative is:

Springer Nature Customer Service Center GmbH
Europaplatz 3
69115 Heidelberg, Germany

**Batch number: 10370708**

Printed by Printforce, the Netherlands